Pipelines: Industry's Lifeline

Furguson

Table of Contents

Chapter 1

Introduction

1.1 Motivation

As a result of strong growth in U.S. and Canadian oil and natural gas production, pipeline use is expected to push capacity to its limit in the future, requiring new pipelines and pipeline expansions to provide access to new markets. According to Natural Resources Canada, there are currently more than 840.000 km of transmission, gathering and distribution pipelines in Canada. These pipelines are aging and increasing demands posed by harsher service conditions stress the importance of integrity management [18]. As such, leakage detection and maintenance are crucial.

Generally, a pipeline is a network of connected pipes with pumps, valves and control devices to help convey liquids or gases. A pipeline network consists of gathering systems, trunk links and distribution systems. The latter system is the longest of the network. Typically, the pipes that form the distribution system are of a small diameter and operate at low pressure.

Pipelines operate year-round and computerized operations are continuously monitoring the pressure, the flow and the consumption energy throughout the line. Software can perform leak detection calculations quickly as well as initiate remedial actions in case of emergency. However, research suggests that operating stations identify only about 15-20 % of the total pipeline leaks [19].

According to the Canadian Energy Pipeline Association, the use of sophisticated technology allows operators to see anything out of the ordinary, for example minute cracks or signs of corrosion, from inside the pipe. Like a small submarine, devices called smart PIGs (Pipe Inspection Gauge) are used for in-pipe inspection. These devices are not autonomous. Rather, they are tethered and move along with the fluid. Currently, the operation of robots for the maintenance of the pipeline utilities is considered one of the most attractive solutions available [20].

1.2 Problem Description

Since pipelines are confined environments, an autonomous robot is a widely accepted method for collecting data, including images of pipes, with little or without any human intervention. Due to the flexibility and adaptability required of the in-pipe robots, several types have been developed. However, many of these robots are tethered robots, therefore their inspection range is limited by the cable length.

Nowadays, there are robots that combine different mechanical and electrical systems to enhance their performance when moving and sensing inside the pipeline. While plenty of in-pipe robots exist, as described in Chapter 2, each has their own limitations and drawbacks. This research will attempt to use a combination of robot morphologies to increase performance. The result of this work can then be integrated with a variety of sensors for trajectory tracking and collection data purposes.

1.3 Objectives

The main objective of this thesis is to develop an in-pipe robot. The proposed robot is comprised of three modules: two propulsive modules and one control module. To this end, a process containing various steps needs to be followed.

As a first step, this research must provide the optimal dimensions such as the radius of the wheels and the distance between shoulder joints in order to integrate all of the necessary components into the propulsive modules. The next step is to design a control module capable of carrying an on-board computer, a battery and the essential electronics to provide locomotion to the robot. A subsequent step is to analyze and adapt the dynamic controller proposed by Douadi *et al.* [17] into the robot's architecture along with simulations that evaluate the controller and the robot's design. Finally, open-loop and closed-loop experimental tests are executed to validate the numerical results.

1.4 Thesis Contribution

The robot's dimensions are optimized in order to move through different pipe diameters. The propulsive module and the control module have been developed. The propulsive module contains all of the actuators and sensors, while the control module carries the required electronics. The adaptation of an existent controller is proposed to generate optimal trajectories during navigation across single elbows.

Simulations and experimental tests have also been performed to evaluate the optimized controller, the selected components, the architecture of the robot, and the general performance of the robot. Recommendations have been made to improve the design of the propulsive module and the control module of the robot for future research.

1.5 Thesis Outline

The remainder of the thesis is organized as follows:

- Chapter 2 presents a review of the architecture of relevant in-pipe robots and sets the framework in which the current robot is situated.

- Chapter 3 describes the proposed mechanical, electrical and software designs of an in-pipe robot for inspecting pipes of different diameters.

- Chapter 4 analyzes an optimized 2D controller for the in-pipe robot and includes simulations in different scenarios.

- Chapter 5 evaluates experimental results obtained by open-loop, closed-loop, and autonomy tests. This analysis establishes the performance of the optimized controller, actuators and sensors of the robot.

- Finally, Chapters 6 and 7 discuss the main experimental results and the architecture of the robot. A few possible future research avenues are suggested.

Chapter 2

Literature Review

This chapter contains a literature review of relevant in-pipe robot design, sensing, control and processing to give the reader an insight into the different technologies that have been developed over the years to solve the issue of inspection in confined spaces such as pipelines. The majority of in-pipe robots can be classified into several elementary categories according to their movement patterns, as shown in the Fig. 2.1. Currently, Canada uses Pipe Inspection Gauge (PIG) robots, which are tethered devices that move along the pipeline with fluid pressure. However, the support type robot is well known among researchers for its adaptability. In the following sections, each morphology is described in the order presented in Fig. 2.1.

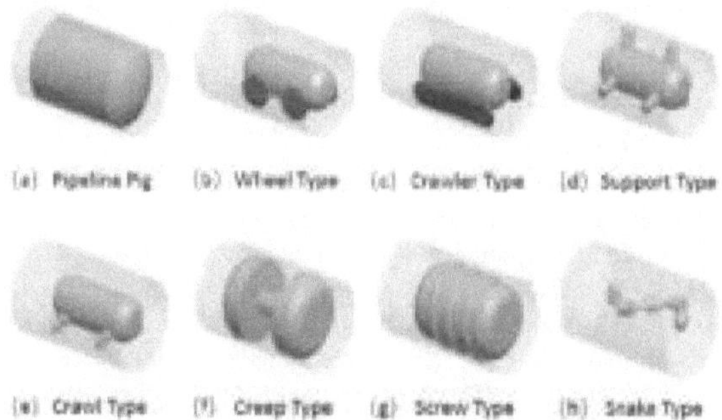

Figure 2.1: Classification of in-pipe robots [1]

2.1 Pipeline PIG

The PIG type robot illustrated in Fig. 2.1(a) is one of the most well-known commercial in-pipe robots; it is a small device that resembles a submarine. The fluid pressure propels the robot along the pipe, preventing it from executing sophisticated movements.

Okamoto *et al.* have presented a PIG robot that is shown in Fig. 2.2. The robot has a cylindrical capsule that is connected to rubber discs by passive joints, which allows the capsule to rotate around its longitudinal axis. The principal purpose of the rubber discs is to block the fluid and to propel the robot; moreover, the discs centre the capsule into the pipeline. This robot has multiple ultrasonic transducers to measure the echo time propagation and to analyze the inner wall of the pipeline. An odometer is used to determine the distance that the robot is travelling [2].

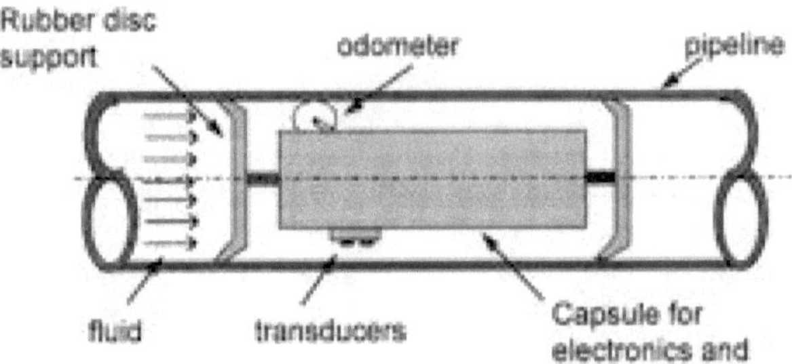

Figure 2.2: PIG robot developed by Okamoto *et al.* [2]

The capsule has an on-board computer, an odometer, ultrasonic transducers and batteries. The computer activates one ultrasonic transducer at a time. When the transducer reading is done, the computer uses an algorithm to determine the occurrence of corrosion in the inner wall. The computer reads the odometer with the same protocol as the transducer but at different frequencies. This frequency avoids interference between signals while saving all of the data. The on-board computer is a 16-bit microcontroller that can be programmed in Forth or in Assembler and operates at the clock frequency of 20 MHz. These languages combine speed and versatility as they are low-level languages.

A set of nickel-cadmium batteries powers the control capsule with a nominal tension of 12 V and a capacity of 7 Ah. The standby current consumption of the microcontroller is of 300 µA and the typical current consumption is of 31 mA. However, the total consumption of the robot is estimated at 2.4 Wh. Therefore, the battery can supply the control capsule for up to 35 h. Fig. 2.3 depicts the block diagram of the electronics aboard the robot.

The on-board computer filters the gained data with an algorithm that saves the defects' location due to the disk's storage space limitation. The external computer and the power supply are connected to the robot by two high-pressure resistant connectors. These connectors are called communication and power connectors, respectively. The first connector uses a serial protocol to communicate with the robot. The serial channel has four

commands: measurement test, defect level, start inspection and end inspection. A code in
C++ receives the data from the serial channel displays it in a graphical interface where
the user can see the defects throughout the pipeline.

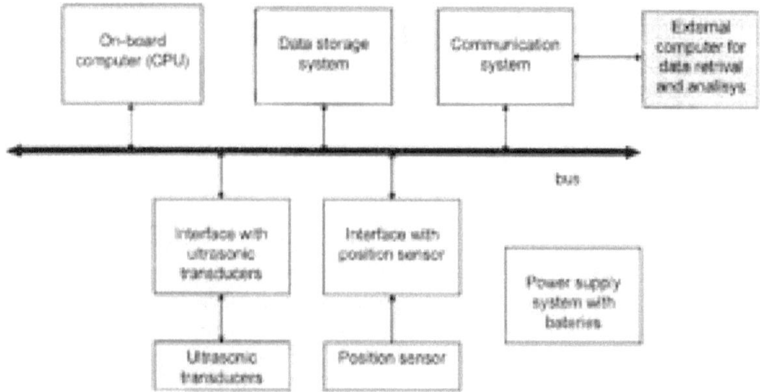

Figure 2.3: Control of the PIG robot developed by Okamoto *et al.* [2]

Hu and Appleton have proposed a self-driving PIG robot [3]. Fig. 2.4 shows the
architecture of the robot. The robot uses a turbine to transform the kinetic energy of
the fluid and propel the robot through the pipe. The direction of the robot's turbine
determines if the robot will move forward or backward, since the shaft is connected to a
reverse-transverse screw. This screw converts the rotational movement into a translational
propulsion force for the robot's motion.

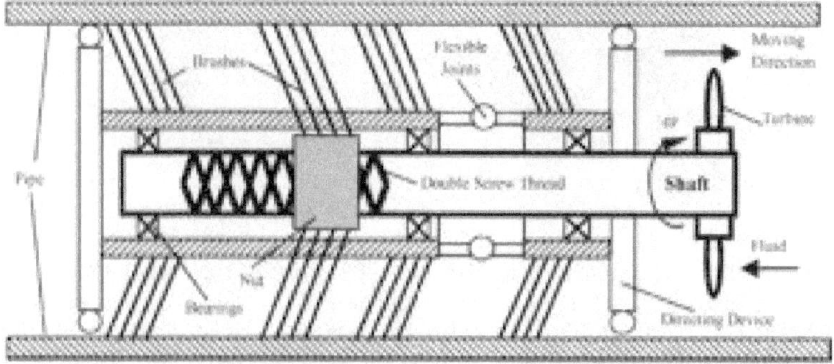

Figure 2.4: PIG robot created by Hu and Appleton [3]

The robot has brushes around its main body and around the nut at the end of the screw. The brushes are in contact with the inner wall of the pipe, and they can slide along the pipe but can also hold the robot's position. Evidently, the robot is completely mechanical and does not include sensors or electronics. Thus, the dynamics of the robot depend on the robot's dimensions, the forces of friction, and the speed and density of the fluid. Experimental tests were developed in an arrangement of pipes with water to propel the robot. Therefore, the maximum speed of the robot depends on the fluid speed, which in this case was $2\,\text{m/s}$.

2.2   Wheel Type

The wheeled type robot, shown in Fig. 2.1(b), is one of the most common practices for inspection in-situ pipe. A cylindrical workspace is the typical representation of round ducts or pipelines, and is a very common geometrical shape [12]. These robots can move smoothly and fairly quickly along horizontal pipes because of the convenience of wheel-based locomotion. However, they cannot operate inside vertical or inclined pipes as a consequence of the lack of support around the pipe. For instance, Song *et al.* developed a controller to guide wheeled mobile robots inside cylindrical workspaces based on a kinematic model [4]. The prototype is a car-like mobile robot with wheels of $50\,\text{mm}$ of radius. The setup used a $500\,\text{mm}$ pipe to test the controller, as is shown in Fig. 2.5. Various experiments, with a robot linear speed of $2\,\text{m/min}$, were conducted. During these tests, they modified the control law parameters to determine the effectiveness of the controller.

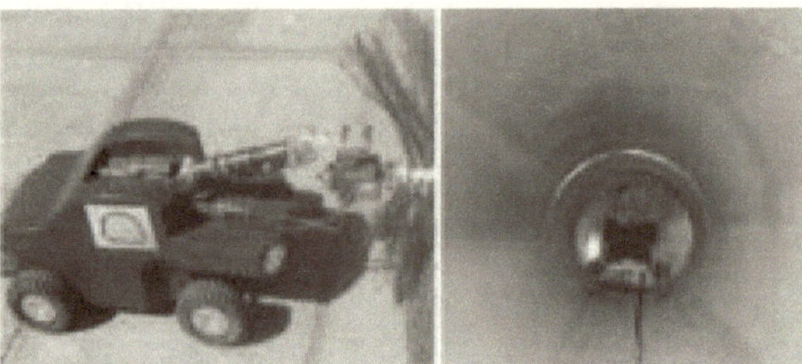

Figure 2.5: Setup to test the car-like robot controller created by Song *et al.* [4]

Kawaguchi *et al.* presented another case of a wheeled type robot [5]. The car-like robot uses a mechanism based on dual magnetic wheels, which allows it to move inside inclined pipes and cope with vertical pipes. However, the magnets limit the robot's motion inside iron and similar pipes by virtue of the magnet's properties. The overall weight and length of the robot is $4.25\,\text{kg}$ and $410\,\text{mm}$, respectively. The robot has two motors in order to propel the rear wheels and the front wheels individually. Fig. 2.6 shows the general architecture of the robot.

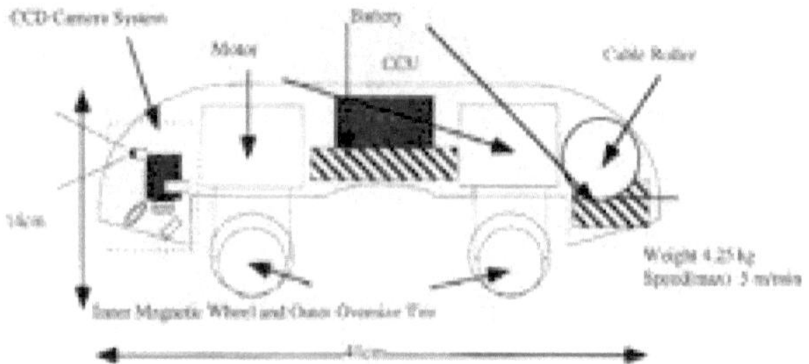

Figure 2.6: General architecture of a wheeled robot developed by Kawaguchi *et al.* [5]

The electronic system of the robot is composed of a CCD camera, a camera control unit, a controller, a motor, as well as O/E and E/O converters for signal reception and transmission. Fig. 2.7 shows the configuration of the electronic system and the principal components. The robot is not completely autonomous because it needs a fibre-optic cable to send and receive information from the computer. However, the quality of the cable allows the robot to move up to 500 m without signal attenuation problems. They placed the CCD camera facing downward and it uses a prism mirror to observe the forward and the backward directions at the same time.

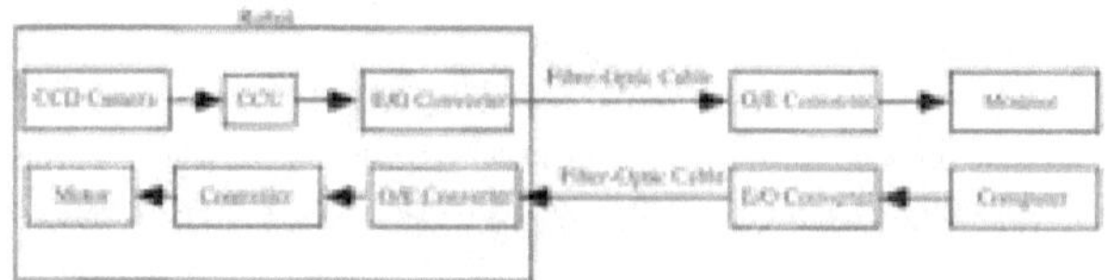

Figure 2.7: Control of the wheeled robot developed by Kawaguchi *et al.* [5]

2.3 Crawler Type

Crawler or Caterpillar type robots are comparable to wheeled robots but are characterized by a higher traction, which can be useful in certain conditions, such as slippery surfaces or inclined pipes. Nonetheless, they still cannot propel themselves in vertical pipes [21]. Kin *et al.* developed a prototype that is somewhere between a crawler and a support type robot [6]. The robot's principal mechanism to ensure adaptability to varying pipe diameters is a modified scissor-lift mechanism driven by pneumatic cylinders. The robot has three position sensors in each cylinder to adjust the pushing forces over the inner wall of the pipe. Moreover, this adjustment allows the vehicle to adapt to pipes with a diameter between 600 mm and 800 mm.

Figure 2.8: Crawler prototype created by Kin *et al.* [6]

The assembled prototype is shown in Fig. 2.8. The total length of the tracked driven module is 680 mm while the prototype weight is 100 kg. Tree transmissions with gears and a DC motor drive the locomotion of the prototype. The robot uses a power and control cable to activate the actuators and flexible plastic tubing to supply air to the cylinders. The maximum speed of the driving module is 2 cm/s. However, this speed requires 21.22 A and an air compressor to work effectively.

2.4 Support Type

Various support in-pipe robots have been studied and developed [22, 23, 8, 24, 25, 26]. These robots can adapt their structure to pipes of various diameters, but they usually need another propulsive module to provide the steering movement inside fittings. The structure of this type of robot is suitable for long-range inspections and carrying heavy loads [27].

The robot presented by Kwon *et al.* can inspect pipelines between 80 mm and 100 mm of diameter [7]. It uses two driving modules connected by a compression spring. The modules are offset by 60° to help with the robot's agility while avoiding motion singularities. Each module has a triangular linkage structure. The linkage has two four-bar structures connected to the hinge joints in the prime body. The caterpillar wheel is made of two gears and a wrapping silicon belt for a large friction coefficient. Fig. 2.9 shows the two modules, linked by a compression spring.

The robot is controlled by three main components: the sensor unit, the power unit and the control unit. The power unit is responsible for moving the actuators according to the needed motion. The sensor unit sends the accelerometer, gyroscope and encoder data to the control unit. The control unit analyzes the data to ascertain the required path. The robot is considered to be a tethered robot as it requires a power/communication cable connected to the computer. Fig. 2.10 shows a flow chart of the robot with its principal components.

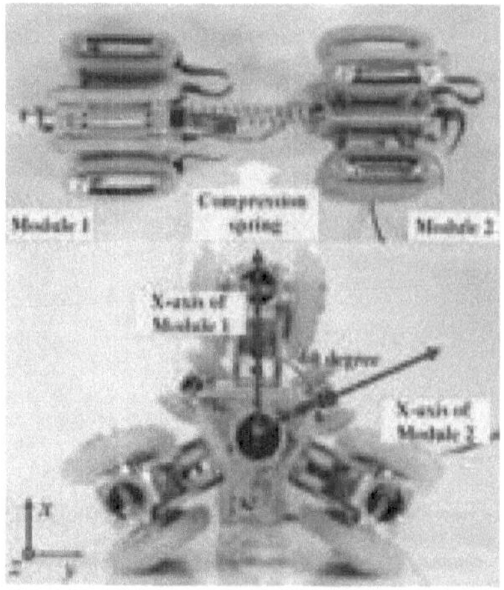

Figure 2.9: General architecture of a support robot created by Kwon *et al.* [7]

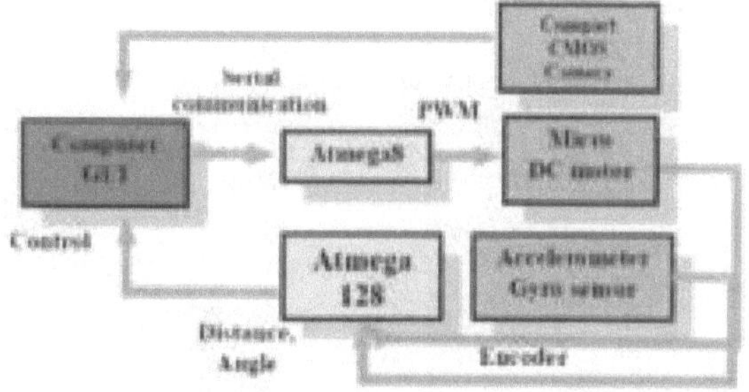

Figure 2.10: Control of the support robot developed by Kwon *et al.* [7]

MRINSPECT IV, a support robot, was developed by Roh and Choi [8]. The robot can be divided into three major parts: a body frame, three driving modules, and a Charge-Coupled Device (CCD) assembly. The main frame has two sets of slider-crank mechanisms to move the driving modules and adjust the body into the pipe and to overcome any irregularities in the inner walls. It is also a holder for the CCD assembly. Each driving module has a geared DC motor with encoder. The motor drives the rear and front wheel using a gear transmission. The CCD assembly is composed of a CCD camera, a light and a sub-wheel mechanism, which prevent the robot from having direct contact with the inner wall of the pipe and also help the robot slide on the inner wall when steering. Fig. 2.11 shows an exploded view of MRINSPECT IV.

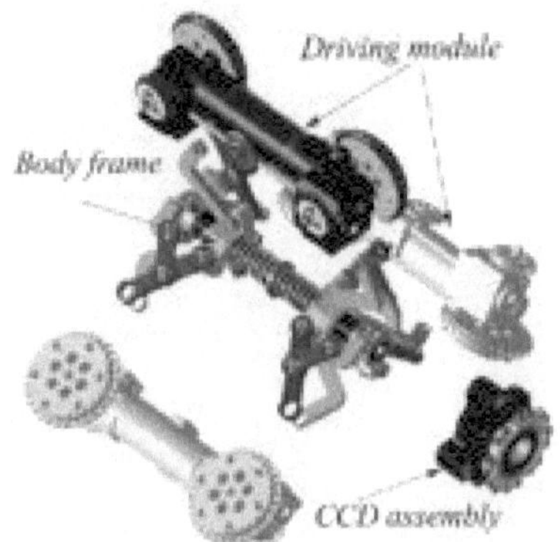

Figure 2.11: Exploded view of MRINSPECT IV developed by Roh and Choi [8]

MRINSPECT IV is a tethered robot controlled by a computer and powered by an external energy supply. The user interface has two sections, the right section shows the data information of the robot and the left section shows the image from the camera. The interface uses a joystick and a keyboard to select between a forward, backward and steering motion on the images from the CCD assembly. Fig. 2.12 depicts the overall control communication.

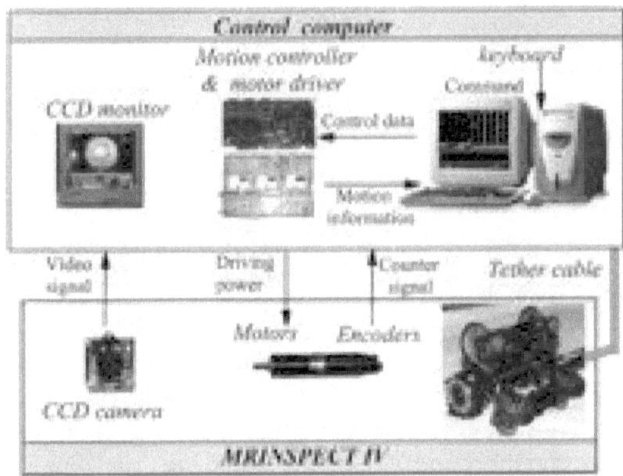

Figure 2.12: Control of MRINSPECT IV developed by Roh and Choi [8]

2.5 Crawl Type

A crawl type in-pipe robot is able to move through straight horizontal pipes and elbow pipes at 45° and 90°. Nevertheless, it cannot move along vertical or inclined pipes. Yu *et al.* presented a walking robot based on planetary gears for horizontal pipes [9]. The prototype is composed of three primary units: a supporting unit, a driving unit and a walking unit. Fig. 2.13 shows the assembled robot.

Figure 2.13: Crawl prototype created by Yu *et al.* [9]

The supporting unit is a platform with four passive wheels, where the platform is linked to the robot by springs. The springs allow the robot to adapt to different pipe diameters. However, the adjustment range of the robot is small. The driving unit has a motor that moves a planetary gear set. This planetary gear set transmits its rotation into two different axes, one on each side of the robot. The walking unit is composed of two legs that are connected to the rotation axes of the driving unit. When a leg is lifting up, the other is in contact with the inner wall of the pipe. The tethered robot has an overall length of 30 mm.

Kejie *et al.* developed an in-pipe robot that is a combination of the crawl and the support types [10]. The tethered robot is capable of moving along horizontal, vertical and bending pipes of diverse diameters. The robot uses a motor to drive the central telescopic mechanism in order to stretch out and compress the structure. This motor is connected to the fore-and-aft self-locking mechanisms to adjust and maintain the robot into the inner wall of the pipe. The robot uses springs to connect the central telescopic mechanism to the supporting mechanism. The mechanical properties of the springs allow the robot to steer and to restore its original position after a turning movement. Fig. 2.14 depicts the overall structure of the proposed robot.

Fig. 2.15 illustrates the motion cycle of the robot. The arrows in the diagram show the movement direction of the robot. The average speed in straight pipelines is 2.7 cm/min and the overall weight of the robot is 0.4 kg. The structure allows the robot to move in pipes of between 90 mm and 150 mm. An external power supply of 12 V powers the tethered robot.

Figure 2.14: General architecture of a crawl robot created by Kejie *et al.* [10]

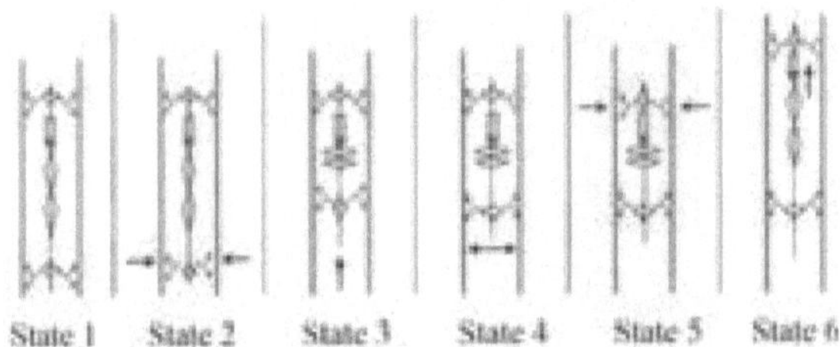

Figure 2.15: Movement cycle of the crawl robot developed by Kejie *et al.* [10]

2.6 Creep Type

A creep type in-pipe robot is reliable for inspection in vertical and inclined pipes due to the force that it applies to the inner wall of the pipe. However, it can more easily get wedged in elbow pipes. This type of robot usually has two units connected by a joint that can rotate freely between the two. Zhang *et al.* presented a squirm pipe robot that uses the creep principle to navigate along the pipe [11]. The robot is composed of three key components: the guide head, the right body and the left body. As the name states, the guide head leads the robot along the pipe. The right body has a flexible helical axle and it crosses the left body. The left body has a gear nut that meshes with the helical axle to generate the movement along all kinds of pipes. The left and right bodies have a set of magnetic wheels with an electromagnetic braking system to fix or release the wheel position of the left or the right parts of the robot. As a result of the magnetic wheels, the robot can only move in metal pipes. Fig. 2.16 illustrates the structure of the robot.

Figure 2.16: General architecture of a creep robot developed by Zhang *et al.* [11]

2.7 Screw Type

The screw drive in-pipe robot requires only one motor to drive within the pipe. This greatly simplifies the mechanical structure and the control system of the robot. For this reason, many researchers have developed this type of robot [28, 29, 30]. Just like creep-type robots, these robots also have the tendency of occasionally getting caught inside the pipe. The wheels on the robot push into the inner wall of the pipe at all times, which allows it to easily climb vertical pipes [31].

Kakogawa *et al.* developed a screw drive in-pipe robot that uses only two actuators to navigate through a bent pipe and a T-branch of 190 mm to 129 mm of diameter [12]. The robot is composed of three major units: front, middle, and rear. The front unit has three arms with angled wheels on the tips. The arms in this unit use a coil spring to ensure the contact of the wheels with the inner wall. The middle unit has two arms with tension springs inside them to adjust the structure in the centre of the pipe. The rear unit has three arms with passive wheels. This unit holds two DC motors, the driving and the steering motor. It links the driving motor to the front unit by a double-unit joint that allows the front unit to turn up to 90°. The steering motor uses gears to transmit the motion to the front unit.

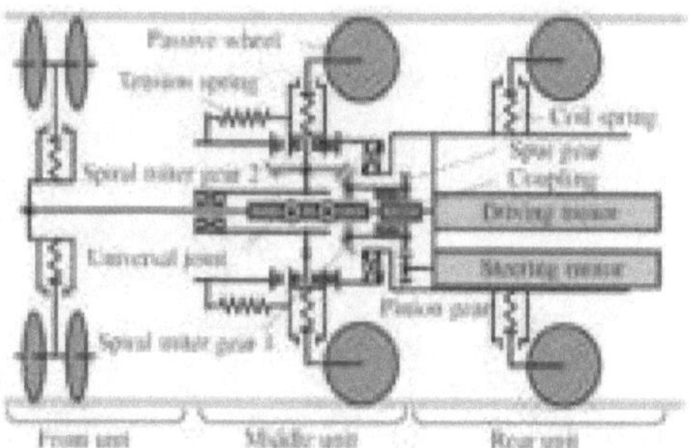

Figure 2.17: General architecture of a screw robot created by Kakogawa *et al.* [12]

The controller mainly consists of a power source, a microcontroller, a motor driver, a mode switch, a joystick and a potentiometer. The joystick moves the robot according to the selected direction. The potentiometer controls the motion speed of the robot. The microcontroller sends signals to the motor driver depending on the received inputs. The mode switch changes between the three-control modes of the robot. Screw driving mode, for forward/backward movement; steering mode, to navigate through branch pipes or elbows; and rolling mode, to change its navigation direction in pipes where it cannot steer. Fig. 2.18 shows the primary components of the controller and their interconnections.

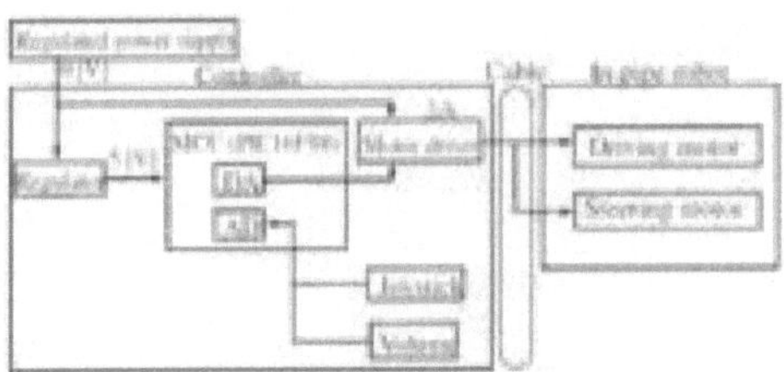

Figure 2.18: Control of the screw robot created by Kakogawa *et al.* [12]

Another screw drive in-pipe robot was developed by Li *et al.* This robot can travel along pipes between 180 mm and 220 mm, at a maximum speed of 7.5 cm/s. It has three principal parts: a suspension mechanism, four driving units and four electromagnetic brake mechanisms. The suspension mechanism uses two crank-slider mechanisms that are pressed by two springs. This suspension allows the robot to adapt to the inner wall of the pipe. Fig. 2.19 shows a driving unit, which is composed of gears, DC motors and wheels. The driving unit transmits the power from the DC motor to the wheel. The electromagnetic brake mechanism is responsible of changing the power transmission of the DC motor to steer or to drive along the pipe. Fig. 2.20. illustrates the architecture of the proposed robot.

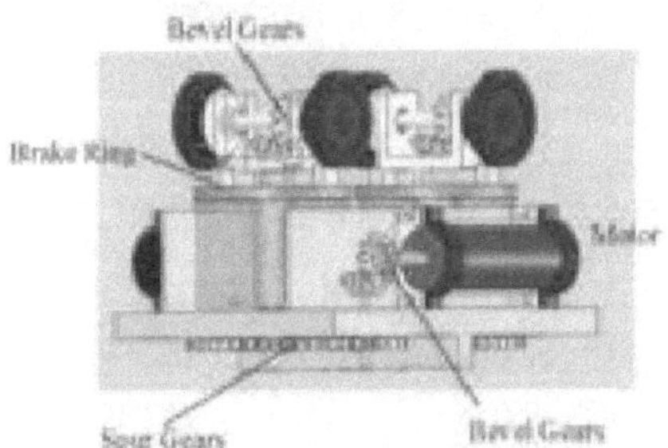

Figure 2.19: Driving unit of a screw robot developed by Li *et al.* [13]

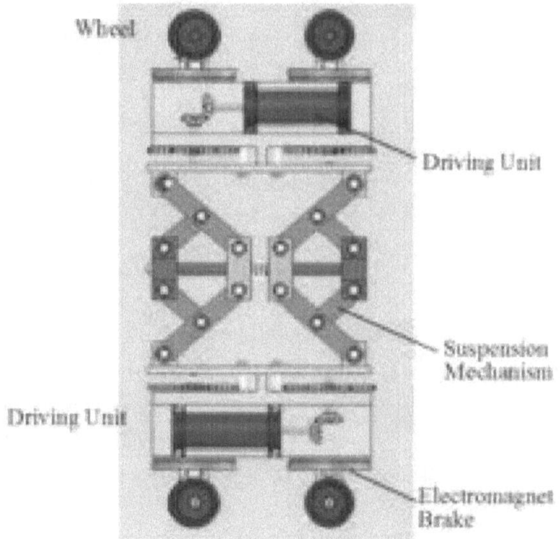

Figure 2.20: General architecture of a screw robot developed by Li *et al.* [13]

2.8 Snake Type

A snake type in-pipe robot has multiple modules with terrain adaptability by means of wheel or foot locomotion. Thus, a snake-like robot has the greatest potential for application in the inspection of industrial pipes [32].

The robot shown in Fig. 2.21 is a spiral pipe-climbing robot that can expand or contract to vary its size to the needed diameter. It has a wheeled multi-module structure, where each module can rotate at the same angle. The rotation between modules is generated by a servomotor and a combination of spur and worm gears. The self-lock characteristic of the worm gear mechanism allows the structure to adapt to the inner wall of the pipe and maintain the desired position. The prototype has two active wheels driven by a DC motor each, as well as two passive wheels to help the robot's motion.

The main structure of the prototype is made of PLA except for the power transmission, which is steel and brass. Based on the kinematics and dynamics of the robot, a controller is programmed in an Arduino UNO R3 to execute the robot's path motion. The robot can move along horizontal, inclined and vertical pipelines.

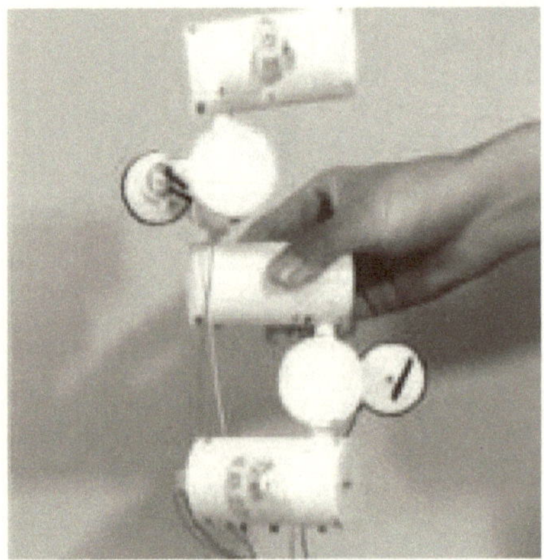

Figure 2.21: SPC robot at 90° created by Dai *et al.* [1]

2.9 Hypermobile Robots

Hypermobile robots are modular articulated mobile robots. According to Robinson and Davis, hypermobile robots can be divided into three groups: discrete, serpentine and continuum [33]. The discrete type is when the architecture of the robot is based on a series of rigid links interconnected by discrete joints. The serpentine type also uses discrete joints but include a short rigid link with a high concentration of joints. The continuum type has an elastic architecture that allows the robot to bend continuously along its trajectory. This type of robot does not have identifiable rigid links or rotational joints.

Usually, discrete hypermobile robots join several wheeled modules into a train to improve speed, load capacity, motion reliability and versatility. The speed and load capacity improvement are caused by the use of active wheels in each module. The motion reliability and versatility are given by the quantity of driven modules, which increases the degrees of freedom. Furthermore, this increase helps the robot overcome any singularities during its course of motion.

Klaassen and Paap developed a hypermobile robot called GMD-SNAKE2 [14]. It consists of six driven modules linked by universal joints, as shown in Fig. 2.22. Each module has active wheels distributed on the outer surface of the module, which are driven by a DC motor. This robot's distribution helps to generate forward force regardless of the module's roll angle. Within the driven module, there are another three DC motors to control the position of the attached universal joint. The joint's position is only limited by the maximum operating angle of the universal joints, which is 45°. The diameter of the robot is 18 cm, the total length is approximately 1.5 m, with a weight of 15 kg.

Figure 2.22: Assembled view of GMD-SNAKE2 created by Klaassen and Paap [14]

The Explorer robot shown in Fig. 2.23, was designed by Schempf and Vradis [15]. It is composed of seven units: two batteries modules, one computer module, two supportive modules, and two locomotive modules. All the modules are linked by articulated joints, which allow the robot to steer in any direction. The locomotive modules have three arms and each arm has an active wheel driven by a DC motor and a gear set. A fisheye camera is placed on both ends of the structure. The supportive module has three arms with passive wheels on the tip of the arm.

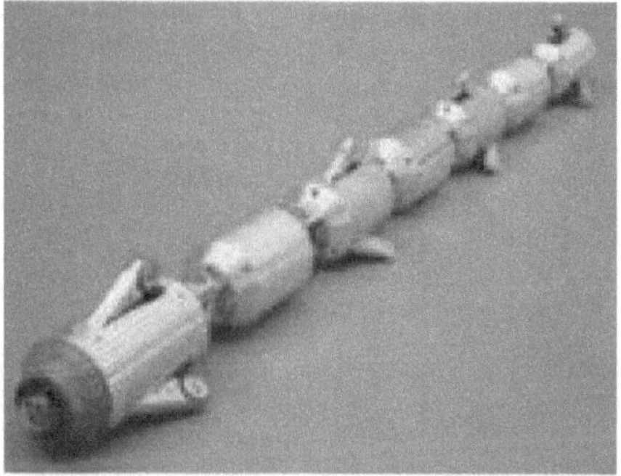

Figure 2.23: Assembled view of Explorer developed by Schempf and Vradis [15]

The robot has a low-power CPU to control the locomotion and steering motion. This process unit uses a set of microcontrollers to manage the inputs and outputs over an internal I2C-bus. Both cameras use a controller to process the video and send it directly to the main board.

2.10 Chapter Summary

In this chapter, the development of in-pipe robots for confined spaces such as pipelines has been discussed. The advantages and disadvantages of each robot presented in this Chapter are summarized in Tables 2.1 and 2.2.

Table 2.1: Comparison between different types of in-pipe inspection robot I

Robot/Type	Advantages	Disadvantages
Okamoto *et al.* PIG	• Simple mechanism • Easily sealed off • Moves in all types of pipes	• Driven by fluid pressure • Tethered • Tendency to get jammed
Hu and Appleton PIG	• Simple mechanism • Easily sealed off • Easy change of direction • Moves in all types of pipes	• Needs a fluid to propel • No sensors • Tendency to get jammed
Song *et al.* Wheel	• High Speed • Moves in horizontal pipes	• Not suitable for vertical and inclined pipes • High tendency of slipping • Tethered
Kawaguchi *et al.* Wheel	• High Speed • Moves in all types of pipes • Suitable for long-range inspection	• Just suitable for metal pipes • High tendency of slipping • Can get stuck if there is a hole in the pipe • Tethered
Kin *et al.* Crawler/Support	• High Speed • Adaptable to a varying pipe • Moves in horizontal and inclined pipes	• Needs air and electricity • High tendency of slipping • Tethered
Kwon *et al.* Support	• High Speed • Adaptable to a varying pipe • Moves in all types of pipes • Suitable for long-range inspection	• High tendency of slipping • Tethered

Table 2.2: Comparison between different types of in-pipe inspection robot II

Robot/Type	Advantages	Disadvantages
MRINSPECT IV Support/Wheel	• High Speed • Adaptable to a varying pipe • Moves in all types of pipes • Suitable for long-range inspection	• High tendency of slipping • Can get stuck if there is a hole in the pipe • Tethered
Yu *et al.* Crawl	• Adaptable to a varying pipe • Moves in horizontal pipes	• High tendency of slipping • Tethered
Kejie *et al.* Crawl/Support	• Adaptable to a varying pipe • Moves in all types of pipes • Simple mechanism	• Leg can get stuck if there is a hole in the pipe • Tethered
Zhang *et al.* Creep	• Moves in all types of pipes • Simple mechanism • Easily sealed off	• Just suitable for metal pipes • High tendency of slipping • Tethered
Kakogawa *et al.* Screw	• Moves in all types of pipes • Simple mechanism • Easily sealed off	• High tendency of slipping • Tethered
Li *et al.* Screw	• Moves in all types of pipes • Simple mechanism • Easily sealed off	• High tendency of slipping • Tethered
Dai *et al.* Snake	• Move in all types of pipes • Adaptable to a varying pipe • Easily sealed off	• High tendency of slipping • Tendency to get jammed • Complex mechanism • Tethered
GMD-SNAKE2 Hypermobile	• Moves in all types of pipes • Easily sealed off	• High tendency of slipping • Complex mechanism • Tethered
Explorer Hypermobile	• Moves in all types of pipes • Easily sealed off • Autonomous	• High tendency of slipping • Complex mechanism

Chapter 3

Proposed Design

This chapter outlines an in-pipe robot design capable of moving along pipes of different diameters. The architecture of the proposed robot is based on the theoretical work of Douadi et al. [34, 17] and the preliminary design of Lamonde [35]. The mechanical components are selected according to the optimal dimensions and the required torques, while the electrical components are selected according to the current and voltage of the actuators as well as the robot's dimensions.

Section 3.1 describes the general architecture and the design guidelines for the proposed robot. Section 3.2 details the mechanical design of the propulsive module. Section 3.3 describes the electronic design of the control module. The programming with block diagrams and calculations is explained in Section 3.4.

3.1 Architecture of the Robot

Knowing the strengths and the drawbacks of the different in-pipe robots presented in Section 2, we have realized that the ideal configuration to provide flexibility during in-situ pipe inspections is a combination of support- and snake-type robots, with independent arm movement and multiple modules. This configuration allows the robot to move along horizontal, vertical, reduction and bent pipes.

The Explorer robot, shown in Section 2.9 [15], is similar to the robot in this research. Table 3.1 shows the main differences, such as the type of universal joints, the type of sensors, and the quantity of modules, arms, and actuators.

Table 3.1: Main differences between the Explorer robot and the proposed robot

Configuration Element	Explorer Robot	Proposed Robot
Universal joints	Actuated	Passive
Sensors	Camera and Encoders	IMU and Encoders
No. of modules	7	3
No. of arms per propulsive module	3	4
No. of actuators	18	16

The steering of the Explorer robot depends on the actuated joints, while the steering of the proposed robot depends on the arms' angle and the motion of the wheels. The Explorer robot has a higher tendency to slip in comparison to the proposed robot because the Explorer robot uses six arms with active wheels to move seven modules, while the proposed robot uses eight arms with active wheels to move three modules. The sensors of the proposed module are used to track the movement of the robot, while the Explorer robot uses its sensors to track the movement but also to display an internal view of the pipe.

The papers of Douadi *et al.* [34, 17] and the thesis of Lamonde [35] focus on a multi-body robot for confined environment inspection. The concept is shown in Fig. 3.1 and consists of identical modules, where each module has independent arms with an active joint. These modules are hitched together by passive revolute joints. This approach does not take into account the physically of the required components to achieve the desired motion. A preliminary investigation by Lamonde [35] determined that a passive free-floating module would be required to house the power pack and the processor.

Taking into account the previous works, the design of a robot to achieve autonomous untethered motion was further developed. The proposed architecture, shown in Fig. 3.2 contains two propulsive modules and one control module. The propulsive modules have four active joints for the rotational displacement of each arm, and every arm has an active wheel that makes contact with the inner wall of the pipe. The control module has an on-board computer, and batteries to power the robot.

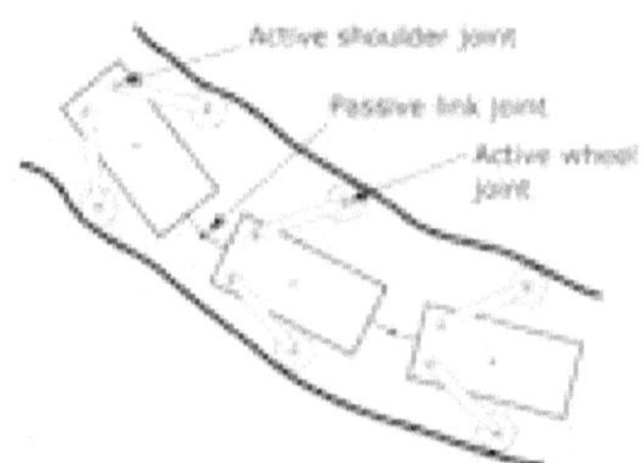

Figure 3.1: Robot concept

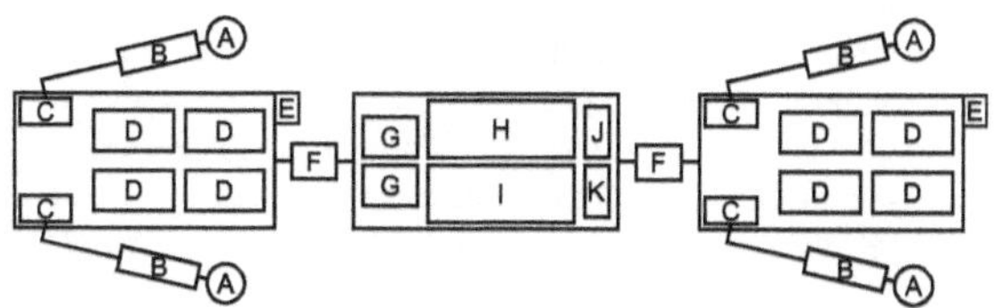

Figure 3.2: Proposed architecture of the robot

Labels from Fig.3.2 are detailed in Table 3.2

Table 3.2: Components of the robot

Label	Component
A	Wheel
B	Gearmotor
C	Worm gear mechanism
D	Digital Servomotor
E	Inertial Measurement Unit
F	Universal Joint
G	Motor Driver
H	Raspberry pi and Servo HAT
I	Batteries
J	Voltage converter for Motors
K	Voltage converter for Processor

The American National Standard for Industrial Robots and Robot Systems defines singularities as "a condition caused by the collinear alignment of two or more robot axes resulting in unpredictable robot motion and velocities". Therefore, a singularity in a confined environment such as a small diameter pipeline is highly likely. According to Douadi *et al.* the proposed robot can present a serial singularity when one or both arms are perpendicular to their respective walls [17]. This scenario is only possible if the robot must travel through a sharp 90° elbow.

The proposed robot architecture adds adaptability and helps to overcome singularities by having two propulsive modules on its ends. If one of the modules encounters a singularity during the trajectory, the robot can still manipulate the other propulsive module to modify its orientation and position. The distance between the shoulder joints, the arms' length, and the wheel radius of each propulsive module is identical, in order to simplify the kinematics and dynamics of the robot.

3.2 Propulsive Module

The propulsive module has two principal mechanisms: a worm drive mechanism to adjust the radial position of the robot within the pipe and a geared drive mechanism to propel the robot along it. The kinematic model of the propulsive module is presented in Fig. 3.3. The geometric variables are the wheel radius ρ, the distance between shoulder joints W, the arm length l, the module diameter D, and the distance between universal joints h. Lamonde [35] created an algorithm to determine the optimal values for ρ, z and W. This algorithm iterates each variable over a specific range and counts the number of collisions produced when the robot navigates inside a pipeline. Nevertheless, due to actuator dimensions, these values are updated in this research to fit into the structure using the most approximate values.

The optimal dimensions are non-dimensionalized with respect to the pipeline diameter and the module length. Since the robot is designed to travel through pipes between 6-inch to 8-inch, it is possible to establish the dimension for the robot based on the smallest pipe diameter.

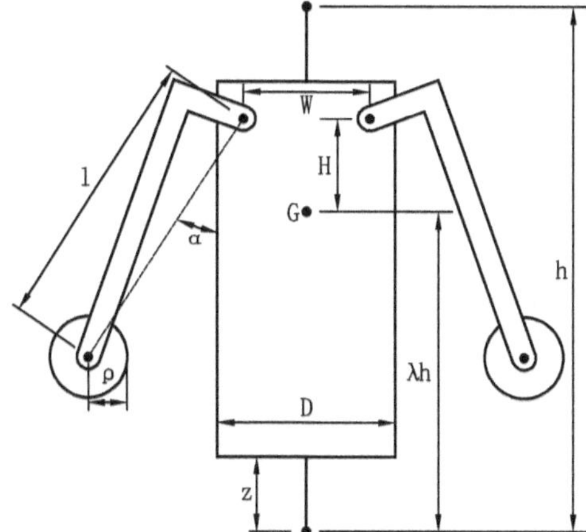

Figure 3.3: Kinematic model of the robot

Table 3.3: Design parameters of the robot

Variable	Parametric Value	Optimal Value (mm)	Actual Value (mm)
ρ	N/A	12.50	15.50
z	N/A	30.00	30.00
W	$0.4\,D_{pipe}$	60.00	51.00
l	$0.75\,D_{pipe}$	114.30	116.00
D	$0.5\,D_{pipe}$	76.20	76.20
h	$D_{pipe} + 2z$	212.00	210.00
λh	$0.65\,h$	137.80	136.50
H	$0.15\,h$	31.80	31.50

3.2.1 Free-Body Diagrams

In this section, two free-body diagrams are presented to calculate the largest of the minimum normal forces. These estimations use the worst-case scenarios that a propulsive module can face during in-situ inspections. Fig. 3.4 shows a horizontal pipeline with three modules, two control modules at the ends and one propulsive module in the middle.

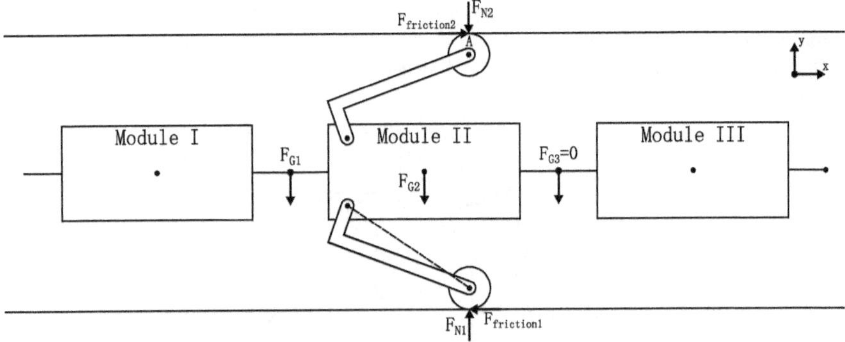

Figure 3.4: Free-body diagram of the worst-case scenario inside a horizontal pipeline

The largest normal force in this scenario is when the gravitational force exerted by the module III is zero, and the arm at the bottom is holding the complete propulsive module while the other three arms are adjusting the module in the centreline of the pipe. This scenario considers that the robot is inside a 6-inch pipe because that is when the weight forces apply the largest torques.

With this free-body diagram, a sum of forces in the y-axis and a sum of moments at point A (contact point in the upper wheel) can be applied:

$$\sum F_y = 0$$

$$F_{N1} - F_{N2} - \frac{F_{G1}}{2} - F_{G2} = 0$$

$$F_{N2} = F_{N1} - \frac{F_{G1}}{2} - F_{G2} \tag{3.1}$$

and,

$$\sum M_A = 0$$

$$F_{friction1}\, l_{f_{friction1}} - \frac{F_{G1}\, l_{F_{G1}}}{2} - F_{G2}\, l_{F_{G2}} = 0$$

$$F_{friction1} = \frac{F_{G1}\, l_{F_{G1}}/2 + F_{G2}\, l_{F_{G2}}}{l_{f_{friction1}}} \tag{3.2}$$

where the weight forces are obtained with the mass of the modules and the gravity force ($F_{Gi} = m_i\, g\ (i = 1, 2)$), the friction forces are calculated with the normal forces times a coefficient of friction between rubber and the pipe ($F_{frictioni} = \mu\, F_{Ni}\ (i = 1, 2)$), and $l_{F_{G1}}$, $l_{F_{G2}}$, and $l_{F_{friction1}}$ are the effective distances from point A to the points of force.

Eq. 3.2 represents the minimal friction required for the propulsive module not to tip. Any increase in a gravity force or any other type of force will cause tipping in the module. Nevertheless, this equation can be rewritten to obtain the normal force in the lower arm:

$$F_{N1} = \frac{F_{G1} \, l_{F_{G1}}/2 + F_{G2} \, l_{F_{G2}}}{\mu \, l_{f_{friction1}}} \tag{3.3}$$

At this point, all the variables in Eq. 3.3 and Eq. 3.1 are unknown. To calculate the forces exerted by the weight of the modules, a mass estimation for the propulsive and control modules can be developed based on the necessary actuators and the kinematic model. Table 3.4 depicts a mass estimation of the principal components in each module. The masses of the actuators, batteries and processor are an average between different components that can fit in the robot design. The arms, structure and control case use the material density to calculate the mass. The arms and the structure are made of aluminum, while the control case is made of PLA. Based on the material densities and the 3D model created in Solidworks, the centre of the mass varies between 60 % and 65 % of the module length, depending on the arms' position.

Table 3.4: Mass estimation of the robot

Module	Component	Mass (gr)	Quantity	Total (gr)
Propulsive	Servomotor	70	4	280
	Gearmotor	15	4	60
	Arm	120	4	480
	Structure	300	1	300
Total				1120
Control	Batteries	100	2	200
	Processor	80	1	80
	Electronics	150	1	150
	Case PLA	350	1	350
Total				780

To calculate the effective distances, the diagram shown in Fig. 3.5 was created. This figure depicts the main dimensions to obtain these distances.

$$Arm_y = \frac{D_{pipe}}{2} - \frac{W}{2} - \rho \tag{3.4}$$

$$Arm_x = \sqrt{l^2 - Arm_y^2} \tag{3.5}$$

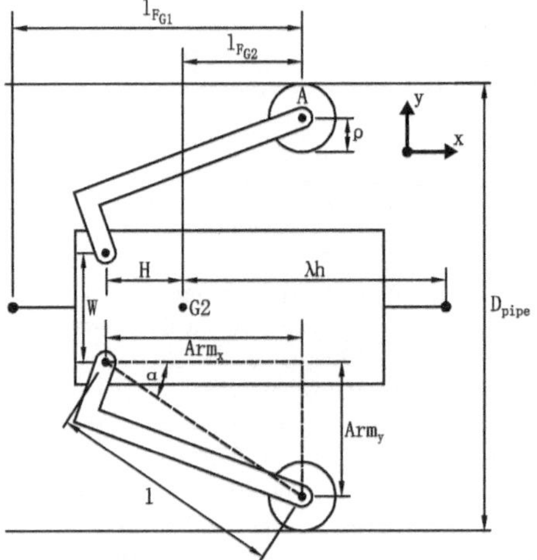

Figure 3.5: Diagram to calculate the effective distances

Therefore, the effective distances can be written as:

$$l_{FG1} = Arm_x - H + h(1 - \lambda) \tag{3.6}$$

$$l_{FG2} = Arm_x - H \tag{3.7}$$

$$l_{friction1} = D_{pipe} \tag{3.8}$$

Knowing the dimensions and the mass of the modules, there are still two unknown variables, D_{pipe} and μ. The nominal pipe diameter is 6 inches but the standard inner diameter is 6.065 inches (See Appendix A). The chosen coefficient of friction between rubber and steel is $\mu = 0.4$ [36]. This coefficient is conservative because the data available does not include a specific value between steel and rubber.

In order to calculate the largest of the minimum normal forces at the bottom wheel, the estimated masses will be multiplied by a security factor in case the selected components are heavier than expected. The propulsive modules use a security factor $SF_p = 1.25$, since the components do not vary a lot in weight. The control module uses a security factor $SF_c = 2.0$, because this module can hold a mixture of batteries and electronics or it can be full of batteries, which will increase the weight drastically. Substituting Eq. 3.6, 3.7, and 3.8 into Eq. 3.3 and Eq. 3.1, the normal forces in this scenario can be calculated as:

$$F_{N1_{min}} = \frac{SF_c\, m_1\, g\, (Arm_x - H + h(1-\lambda))/2 + SF_p\, m_2\, g\, (Arm_x - H)}{\mu\, D_{pipe}} \tag{3.9}$$

$$F_{N2_{min}} = F_{N1} - \frac{SF_c\, m_1\, g}{2} - SF_p\, m_2\, g \tag{3.10}$$

The equations were solved in Matlab (See Appendix B) and the results for the minimum
normal forces in the horizontal pipe scenario are:

$$F_{N1_{min}} = 36.46N$$
$$F_{N2_{min}} = 15.07N$$

A second free-body diagram that depicts the worst-case scenario in a vertical pipeline
is shown in Fig. 3.6. The minimum normal force in this diagram is when the friction forces
of the four arms are equal to the gravitational force exerted by the second module weight
and the half of the weight of the other two modules.

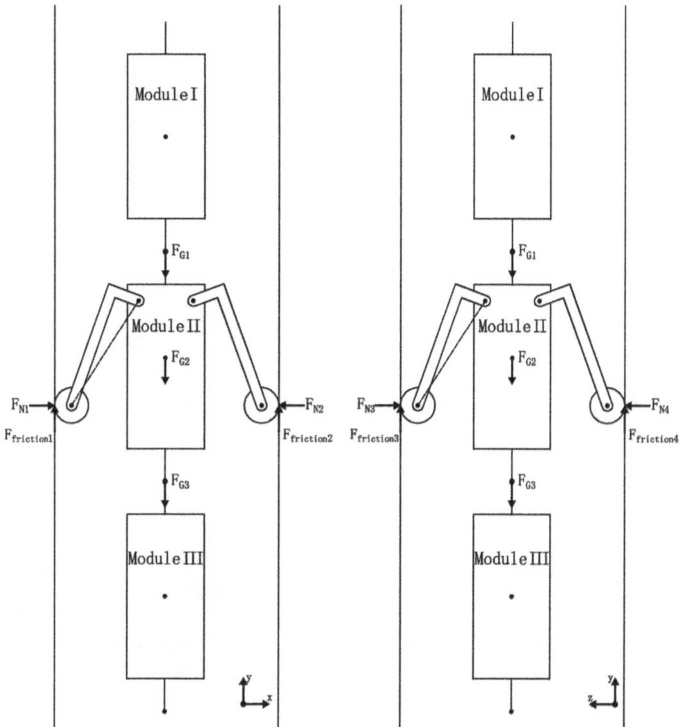

Figure 3.6: Free-body diagram of the worst-case scenario inside a vertical pipeline

Using the x-y plane and the y-z plane, a sum of the forces in the y-axis can be applied:

$$\sum F_y = 0$$

$$\frac{F_{G1}}{2} - F_{G2} - \frac{F_{G3}}{2} + F_{friction1} + F_{friciton2} + F_{friction3} + F_{friciton4} = 0 \tag{3.11}$$

where the weight and friction forces are obtained with the same procedure applied in the horizontal free-body diagram. Thus, $F_{Gi} = m_i\,g$ ($i = 1, 2, 3$), and $F_{frictionj} = \mu\,F_{Nj}$ ($j = 1, 2, 3, 4$).

Assuming that the arms are exerting the same force into the wall and that the robot is centred with the pipeline, the normal forces of all the arms must have the same value. Thus, Eq. 3.11 can be rewritten as:

$$4\,\mu\,F_{N1} = \frac{m_1\,g}{2} + m_2\,g + \frac{m_3\,g}{2} \tag{3.12}$$

Adding the estimated security factors from the horizontal free-body diagram into Eq. 3.12, the minimum normal force in the vertical scenario can be estimated as follows:

$$F_{N1} = \frac{{}^{SF_c}m_1\,g/2 + SF_p\,m_2\,g + {}^{SF_c}m_3\,g/2}{4\,\mu} \tag{3.13}$$

Eq. 3.13 was solved in Matlab (See Appendix B) and the minimum normal force for this scenario is of 17.65 N m. This normal force is smaller than the normal force calculated in the horizontal pipeline scenario.

3.2.2   Worm Drive Mechanism

The worm drive mechanism controls the aperture of the arm with the purpose of keeping the robot in the centreline of the pipe. This motion is executed with a worm screw meshed with a worm gear. Such a mechanism allows a high torque transmission. Although it has low efficiency as a result of the sliding and rolling actions that happen between the worm screw and the worm gear [37], the mechanism is only expected to be operated when a geometry change occurs (e.g. elbows or reduction pipes). The required motion in the arm joints is of 13.25°, it varies between 18.1° and 31.35°, depending on the nominal pipe size (See Appendix C).

Thus, a digital servomotor was chosen for this mechanism. This type of actuator has a microcontroller with a closed-loop algorithm to provide a consistent torque and a smoother motion. Metallic internal gears are used to reach high torques but the standard motion is of 180°.

Each propulsive module has four worm drive mechanisms, two in the x-axis and another two in the y-axis. In order to fit the four mechanisms in the propulsive module, the distance between shaft gears in each worm drive mechanism cannot be higher than 20 mm. The second criterion is that the reduction ratio should be at least 1:10, to ensure that the arm's motion is higher than the required motion of 13.25°. The most suitable option based on these criteria is a worm drive mechanism from KKWG that has a distance between shafts of 10 mm and a reduction ratio of 1:10. This mechanism allows a maximum arm motion of 18°.

The selection of the servomotor is based on the stall torque generated on the shoulder joint. The actuator must deliver enough torque to avoid any slippage on the inner wall of the pipe and it must be able to push the complete body against the pipeline wall when an arm is below the module. Fig. 3.7 shows the torques at the bottom arm with the normal force that was previously calculated.

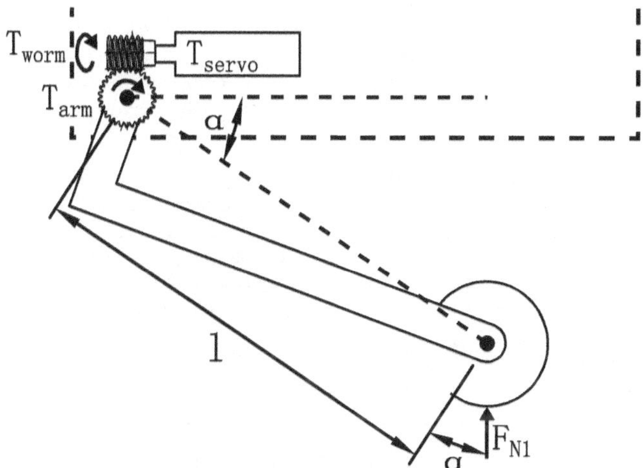

Figure 3.7: Torques and forces for one arm

Based on the normal force and the torques in the worm drive mechanism, three equations can be defined, as follows:

$$T_{worm} = T_{servo}\, \eta_{servo} \tag{3.14}$$

$$T_{gear} = T_{worm}\, \eta_{worm}\, n_{worm} \tag{3.15}$$

$$F_{N1} = \frac{T_{gear}}{l} \cos \alpha \tag{3.16}$$

where the servomotor efficiency is estimated at $\eta_{servo} = 0.9$, the worm drive efficiency is estimated at $\eta_{worm} = 0.5$, and the worm gears ratio is $n_{worm} = 10$.

By collecting Eqs. 3.14, 3.15, 3.16 we can write the servomotor torque as:

$$T_{servo} = \frac{F_{N1}\, l}{\eta_{servo}\, \eta_{worm}\, n_{worm}} \left(\frac{1}{\cos\alpha}\right)$$

(3.17)

$$T_{servo} = \frac{36.46 * 0.116}{0.9 * 0.5 * 10} \left(\frac{1}{\cos\alpha}\right) = 0.94\frac{1}{\cos\alpha}$$

where the maximum value of $\dfrac{1}{\cos\alpha} = 1.36$, because the arm is able to move up to 43° (see Appendix D)

Thus, the minimum torque required by the servomotors is of 1.27 N m. The selected servomotor is the EzRobot heavy-duty servomotor as it has a stall torque of 1.5 N m at 6 V or 1.85 N m at 7.4 V, and a working motion of 180°. These specifications satisfy the required servomotor angle to move the robot through pipelines with diameters between 6 and 8 inches. Moreover, the servomotor stall torque with the self-locking worm gears can maintain the robot's stability in the centreline even with the largest normal force calculated in the free-body diagrams.

Based on the normal forces obtained in Section 3.2.1 and the Eq. 3.17, an operation table can be derived according to different orientations inside the pipeline. Table 3.5 depicts the torques that the servomotors need to exert, as well as the normal forces caused by these torques. This table assumes that the robot is in perfect equilibrium inside a horizontal pipe, which means that the robot is centred in the pipeline. Fig. 3.8 illustrates the orientation of the robot at different angles, with the proper labelling to identify each arm in the propulsive module.

Table 3.5: Normal forces and servomotor torques depending on the robot's orientation

Roll angle	Arm A		Arm B		Arm C		Arm D	
	F_N (N)	T (Nm)	F_N (N)	T (Nm)	F_N (N)	T (Nm)	F_N (N)	T (Nm)
0°	36.1	1.27	0	0	14.72	0.51	0	0
30°	31.26	1.09	7.36	0.25	12.74	0.44	18.05	0.63
45°	25.52	0.89	10.40	0.36	10.40	0.36	25.52	0.89
60°	18.05	0.63	12.74	0.44	7.36	0.25	31.26	1.09
90°	0	0	14.72	0.51	0	0	36.1	1.27

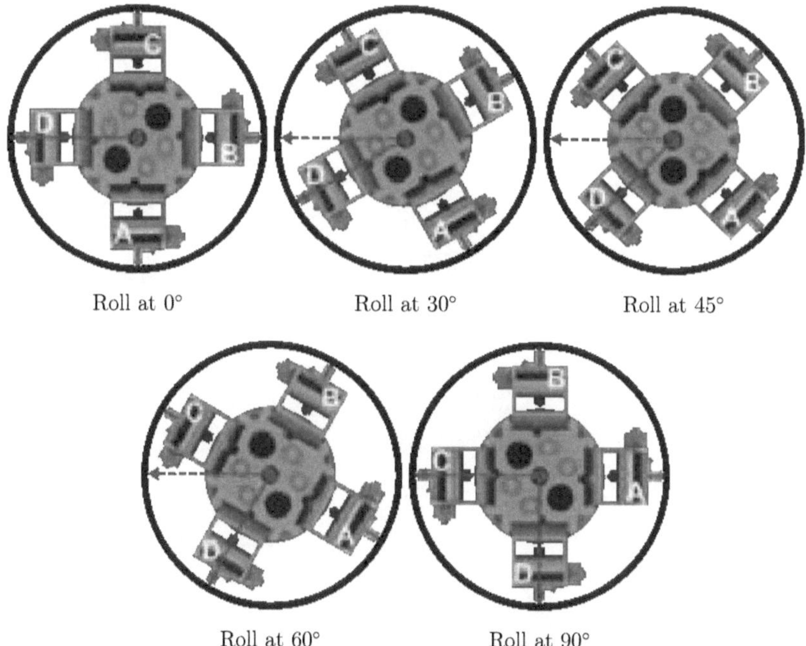

Figure 3.8: Orientation of the robot at different roll angles

Knowing the dimensions of the servomotor and the worm gear mechanism, holder plates for the servomotors and the worm gears were created (See Appendix E). The distance between servomotors must be higher than 6 mm, which is the worm shaft diameter, because two servomotors must be placed at the bottom, and the worm shafts must go through the body. The distance between shoulder joints after a designing process is of 51 mm which, according to Lamonde[35], is in the operative range to avoid slipping in the inner walls of the pipeline.

An exploded view of the propulsive module's internal frame, and a worm gear mechanism for a single arm, is shown in Fig. 3.9. The worm screw plates and the lower plate were machined in aluminum to avoid deflection in the structure. The servomotor plates are identical for an easy assembling process, and they have slots for wire management. Standoffs are used between the servomotor plates to keep the correct distance between levels.

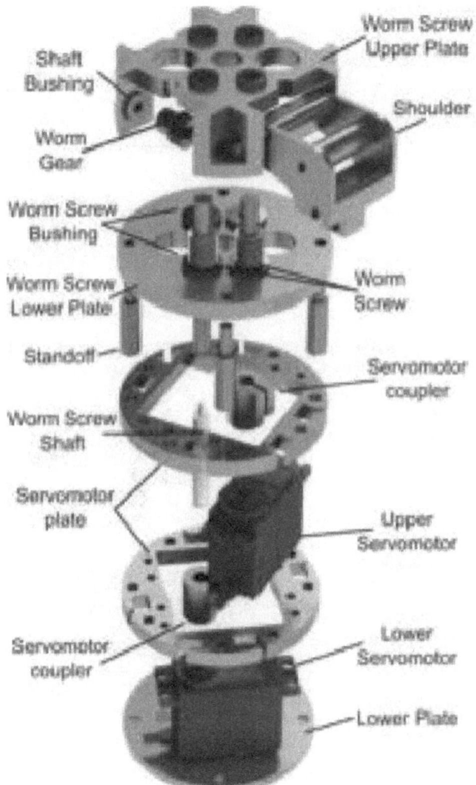

Figure 3.9: Worm mechanism explosion

3.2.3 Geared Drive Mechanism

The geared drive mechanism, which each arm has, provides the propulsion force that the robot needs to navigate through the pipeline. The arms use a right angle driver, which allows them to place the gearmotor along the arm axis. These drivers guarantee that the wheels are in contact with the inner wall of the pipeline at all times. Thus, a reduction set of bevel gear is necessary. The gear ratio was chosen to reduce the torque requirements while providing the needed range of motion, but also to fit inside the arm's dimensions.

A 1:2 bevel gear ratio was selected so that the gear fixed to the wheel would not interfere with the inner pipe wall. The selection of the motor is based on the minimum torque required by the wheel in a horizontal pipe, since the normal force is larger in this scenario (See Section 3.2.1). Assuming that the arms are exerting enough force into the wall to avoid slips, the minimum required torque in each wheel must be equal to the torque that keeps the propulsive module centred in the pipe. Fig. 3.10 shows the forces and torques on the wheel in this scenario.

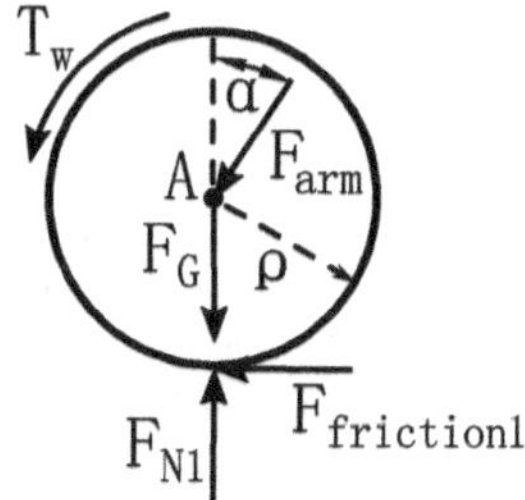

Figure 3.10: Torques and forces on a wheel in a horizontal pipe

Thus, a sum of moments in point A (wheel centre) can be applied to obtain the wheel torque:

$$\sum M_A = 0$$
$$T_w = \rho\, F_{friction1} \tag{3.18}$$

where the friction force is calculated with the normal force times the coefficient of friction between rubber and the pipe ($F_{friction1} = \mu\, F_{N1}$), and the wheel radius is $\rho = 0.0155$.

Using the largest normal force in the worst-case scenario inside a horizontal pipe, the minimum wheel torque is estimated at $T_w = 0.22\,\mathrm{N\,m}$. The wheel torque estimation can also be obtained with the dynamic model presented by Douadi *et al.* [17], where the torque is estimated at $T_w = 0.45\,\mathrm{N\,m}$.

Based on the wheel torque, the following equation expresses the minimum required peak torque of the motor:

$$T_w = T_{motor} \, \eta_{bevel} \, n_{bevel}$$
$$T_{motor} = T_w * \frac{1}{n_{bevel}} * \frac{1}{\eta_{bevel}} \tag{3.19}$$

where the bevel gear efficiency is estimated at $\eta_{bevel} = 0.95$, the bevel gear ratio is $n_{bevel} = 2$, and the maximum wheel torque is estimated at $T_w = 0.45$.

The calculation for the motor selection uses the value obtained from the dynamic model, since it is larger than the value obtained from the free-body diagram. Thus, using Eq. 3.19, the minimum required torque for the motor is estimated at 0.23 Nm.

The selected motor is a Pololu 1000:1 micro metal gearmotor with an extended motor shaft. This component has a stall torque of 0.8 N m at 6 V, and a maximum speed of 32 rpm. These specifications meet the torque requirements needed to move the robot through a vertical pipeline but limit the speed of the robot.

Previous studies of this robot determined that a torsion spring was necessary on the shoulder to help the arm's motion [38]. However, this spring is no longer needed as the selected servomotor exerts enough torque to support the propulsive module with two control modules, in the worst-case scenario. The new shoulder was nonetheless designed to hold the torsion spring in case it is needed in a future version of the robot.

An exploded view of the geared drive mechanism for a single arm is shown in Fig. 3.11. The wheels use a metrical o-ring 2.5x22 made of Buna N. The wheel profile was designed to ensure that the contact points of the wheels are positioned along the longitudinal axis of the propulsive module. The shoulder is composed of two parts to facilitate the assembling of the worm gear and the spring. The torsion spring must be placed around the spring hub on the shoulder.

The arm profile is a connection between the shoulder and the arm holder, the latter being hollow in order to decrease the total weight of the robot. Since stresses on this component can bend the profile, stress calculations were performed to ensure the profile would not fail. This analysis is presented in Appendix F, where the maximum equivalent stress is 120 MPa and the yield strength for aluminum is 241 MPa, meaning that the arm is not going to be affected by the various stresses.

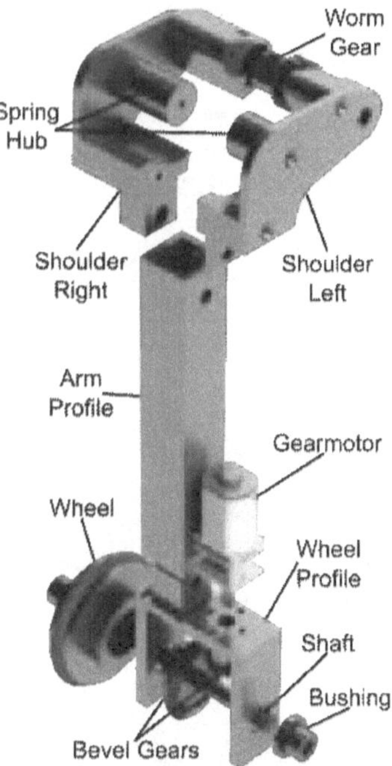

Figure 3.11: Arm explosion

3.3 Control Module

The control module is placed between two propulsive modules. The components inside the module can vary depending on the purpose of the inspection. The elements of the control module developed for this work are those required to evaluate the navigation capabilities of the robot. These are illustrated in Fig. 3.12

There are two main sensors to keep track of the robot's motion, the inclination sensor and the rotation sensor. The inclination sensors measure the tilt angle of the propulsive modules and the rotation sensors measure the angular position of the gearmotors. The robot's navigation is executed by sixteen actuators, eight servomotors and eight gearmotors. Therefore, motor drivers are required between the on-board computer and the actuators in order to control the speed, position, and direction of all the actuators at the same time.

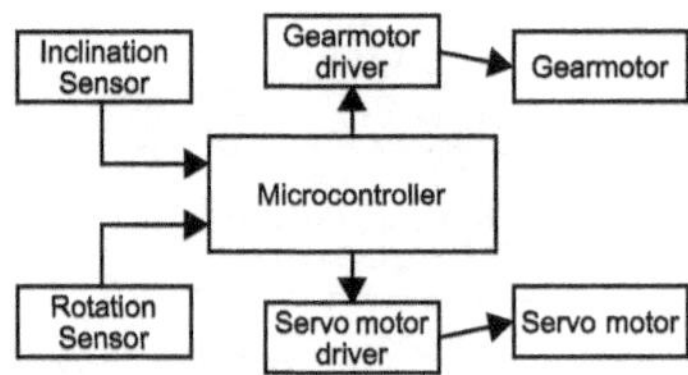

Figure 3.12: General electrical diagram of the robot

3.3.1 Control

An autonomous in-pipe inspection robot must navigate, collect data, and generate reports about the anomalies along the pipe surface. Therefore, the microprocessor should be able to establish a bidirectional wireless communication to display a graphical user interface (GUI).

This microprocessor must also control all the actuators and sensors of the robot. Table 3.6 shows the main characteristics between the two principal microprocessors considered, an Arduino Mega and a Raspberry Pi 3 B+.

Table 3.6: Comparative between Arduino Mega and Raspberry Pi 3 B+

Characteristic	Raspberry Pi 3 B+	Arduino Mega
CPU	Quad cortex A53 @1.2GHz	ATmega 2560 @ 16 MHz
Flash	External SD (up to 256 GB)	256 kB
RAM	1 GB SDRAM	8 kB SRAM
Ethernet	10/100	NA
Wireless	802.11n/Bluetooth 4.0	NA
Current	230 mA	50 mA
Voltage	5 V	31.50
Size	85 x 49 mm	101.6 x 50.8 mm
Operative System	Raspbian	Arduino IDE

The main difference is that Arduino is a microcontroller, which means that it runs only one program repetitively, while Raspberry Pi is a mini computer that is based on Linux and can run multiple programs at a time. The programming languages in Raspberry Pi are python, C and C++ while Arduino uses C/C++. The overall dimensions of Arduino are slightly bigger than those of Raspberry Pi, and size is a very important factor in order to fit all the components inside the module. Raspberry Pi has a wireless LAN and Bluetooth connection on board, which is primordial for connecting with an external computer.

The only drawback of Raspberry Pi is that the minimal energy consumption is of 230 mA, while for Arduino it is only 50 mA. After taking into consideration all of these characteristics, the Raspberry Pi 3 B+ microprocessor was selected for its versatility and scalability.

The on-board computer uses external motor drivers to control the direction and the speed of the gearmotor. These drivers must support the current consumption of each gearmotor of 1.6 A at stall and they must work at 6 V. Thus, the selected motor driver is a DRV8838, which has a continuous output current of 1.7 A and an operating voltage of 0 V to 11 V. The only disadvantage is that this board is for a single gearmotor, which means that the control module needs eight drivers to control the motors independently.

The servomotors also need a driver to be controlled by the Raspberry Pi. Since the servomotor has an operating voltage of 5 V to 8.4 V, the desired driver must support this voltage. The chosen component is a Servo HAT, powered by a PCA9685. This shield is specially made for Raspberry Pi and it is controlled with two pins through an I2C protocol, which can easily manage up to 16 servomotors at the same time.

The only inconvenience is that the servomotor board does not allow more than 6 V, which means that the servomotor will not reach its maximum speed and torque. However, there is no commercial shield for Raspberry Pi able to support more than 6 V.

3.3.2 Sensing

The control module of the robot uses encoders as rotation sensors and Inertial Measurement Units (IMUs) as inclination sensors. The selected rotation sensor is fully compatible with the Pololu 1000:1 micro metal gearmotor and it is recommended by the motor manufacturer. This sensor board is powered by a TLE4946-2K, which is an incremental hall effect encoder. This type of encoder must use an external microprocessor to save the angular position of the motor.

These encoders provide a resolution of twelve counts per revolution without the reduction gearbox. Thus, the counts for each rotation of the wheels can be calculated with the following equation:

$$Wheel_{cpr} = n_{motor}\, n_{bevel}\, CPR \tag{3.20}$$

where the counts per revolution of the encoder is $CPR = 12$, the ratio of the bevel gears is $n_{bevel} = 2$, and the ratio of the gearbox is $n_{motor} = 986.41$.

$$Wheel_{cpr} = 986.41 * 2 * 12 = 23673.84$$

The previous result must be rounded to the nearest integer since the on-board computer counts integers. Then, every time the Raspberry Pi counts 23674, it is equal to one complete turn of the wheel.

The orientation of the propulsive modules is fundamental for the trajectory tracking of the robot and it is achieved with IMUs. This board must have at least a 3-axis accelerometer to estimate the angular position of the robot with respect to the world frame coordinate. The MPU6050 was selected because it has a dual 3-axis gyroscope and accelerometer (6 axes in total). Moreover, this chip has an in-built option to configure up to two sensors with different physical addresses. This configuration is needed to link the orientation of the two propulsive modules with the on-board computer through I2C.

3.3.3 Power

To reach full autonomy, the robot needs a battery with enough power to control the sixteen actuators for extended periods of time. All of the electrical components work at 5 V or 6 V, which means that the necessary battery should contain at least two cells to provide 7.4 V. However, the width of the batteries cannot be larger than the module diameter, which is 76.20 mm. Therefore, the selected power supply is two 1500 mA h 2S 25-50C lithium-polymer batteries connected in parallel to provide 3000 mA h with an overall dimension of 87 x 66 x 15 mm.

The servomotors, gearmotors, drivers and encoders support up to 6 V, which implies that the system needs a voltage regulator to step down the battery voltage from 7.4 V to 6 V. The electric circuit is connected in parallel between all the electric components to provide the same voltage to all the devices. Then, the sum of the current going through each device is equal to the minimum current that the voltage regulator must supply. Table 3.7 depicts the continuous current of each component of the robot that works at 6 V.

Table 3.7: Continuous current of the devices at 6 V

Component	Continuous Current (mA)	Quantity	Total Current (mA)
Servomotor	1200	8	9600
Gearmotor	70	8	560
Servo HAT	400	1	400
Motor Driver	1.5	8	12
Encoder	6	8	48
Total Current			10620

Consequently, the selected voltage regulator is a D24V150F6. This device is a 6 V step-down voltage regulator and its typical maximum continuous output current is of 20 A, which is larger than the required current of 10.62 A.

There is a need for another voltage regulator because the Raspberry Pi 3 B+ and the IMU boards operate at 5 V. The power consumption of Raspberry Pi 3 B+ is much larger than the power consumption of the IMU that is less than 5 mA. Therefore, the voltage regulator can be selected according to the power supply specifications of Raspberry Pi 3 B+, which is at least 2.5 A. Consequently, the chosen device is a LM22676, a 5 V step-down voltage regulator with a maximum continuous output current of 3 A.

3.3.4 Electrical Circuits

The block diagram shown in Fig. 3.13 illustrates the electrical components aboard the robot along with the main interconnections. The quantity of electrical components in the control module limits the placement and selection of components. Thus, two identical custom motor driver boards were developed, each board gathering all of the inputs/outputs (direction, speed, encoder) of the gearmotors in one propulsive module.

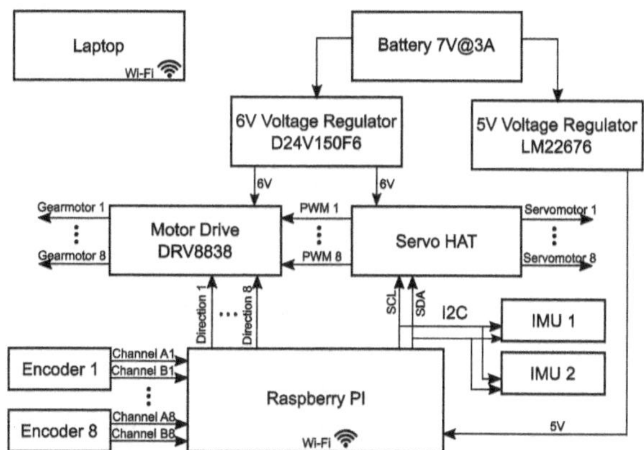

Figure 3.13: Electrical diagram of the robot

A custom circuit board was developed in Eagle, based on the pinout of the drivers and the encoder (See Appendix G). Fig. 3.14 shows the final circuit board and Fig. 3.15 depicts the distribution of each connector. Table 3.8 states the meaning of each label and what pins it contains. The labels use A, B, C, and D to represent the arms in each propulsive module. This board uses JST connectors that can be connected in just one direction to avoid short circuits.

Figure 3.14: Board for motors

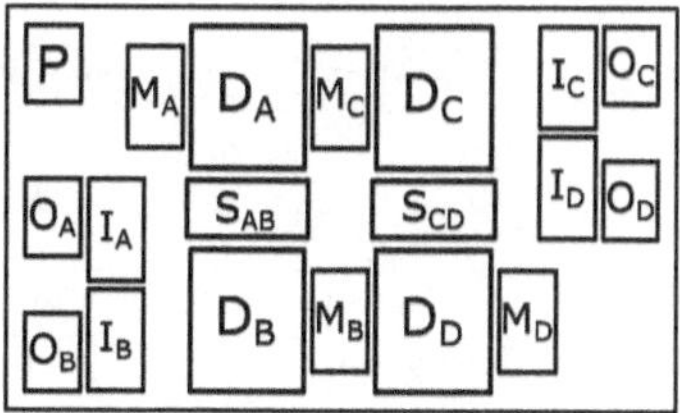

Figure 3.15: Label in board for motors

Table 3.8: Label definition for motor circuit board

Label	Description	Pins
P	Power supply	5V | GND
D_k	Motor driver	N/A
M_k	Motor	VCC | Out2 | Out1
S_k	Signal for the drivers	Phase k | Enable k
I_k	Inputs of encoders	Out A | Out B | GND
O_k	Outputs of encoders	Out A | Out B

k=A,B,C,D

The labels A, B, C, and D were implemented to facilitate the identification of the electrical connections and also to identify the actuators in the user interface. Fig. 3.16 shows a propulsive module seen from above, where the arm at the bottom is called arm "A". The electrical inputs/outputs of each arm are depicted in Table 3.9.

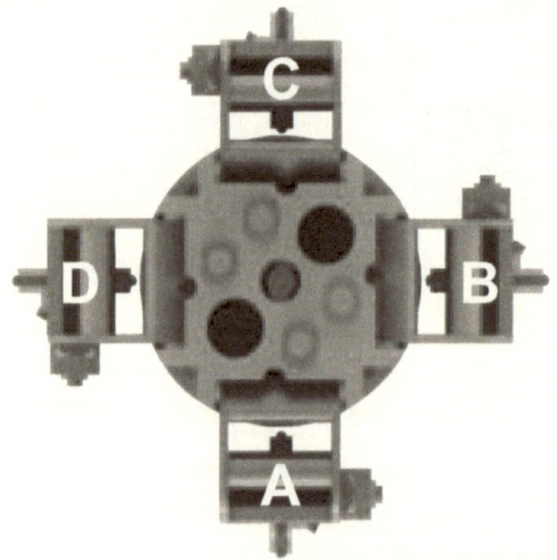

Figure 3.16: Upper view of the propulsive modules with labelling

Table 3.9: Pins description of each arm

Actuator	Variable	Type	Pins
Servomotor	Arm's position	Input	GND | 6V | PWM
	Wheel's direction	Input	Phase
Gearmotor	Wheel's speed	Input	Enable
	Wheel's position	Output	Channel B | Channel A

The Servo HAT has a 2x20 socket header to stock it over Raspberry Pi without wiring connections. There are two IMUs in the robot, one per propulsive module. The Servo HAT and the IMUs use the I2C protocol to communicate with the on-board computer. This protocol can establish a communication with multiple slaves using just two wires, SDA and SCL. The serial data wire (SDA) is used by the master and the slave to send and receive data, while the serial clock wire (SCL) is used as a clock signal to synchronize all the slaves with the master. This protocol uses addresses to communicate with each slave. These addresses are in hexadecimal format and they are usually chosen by hardware on the board. The Servo HAT address is 0x40, the IMU in the back propulsive module is 0x68 and the IMU in the front propulsive module is 0x69.

Raspberry Pi 3 B+ has 40 pins but just 28 are controllable. Two of those pins (SDA/SCL) are used for the I2C communication, sixteen pins are declared as digital inputs, and eight pins are digital outputs. The inputs are used to read the value of each encoder, and the outputs are used to change the rotational direction of each gearmotor. Nevertheless, the motor drivers need two inputs to work, the direction and the speed. The direction is given directly from Raspberry Pi, while the speed is obtained from the Servo HAT. Based on the previous pin configurations of each electrical device, a diagram showing the connection to Raspberry Pi and Servo HAT was created. Fig. 3.17 depicts the specific interconnections between actuators, drivers, and sensors.

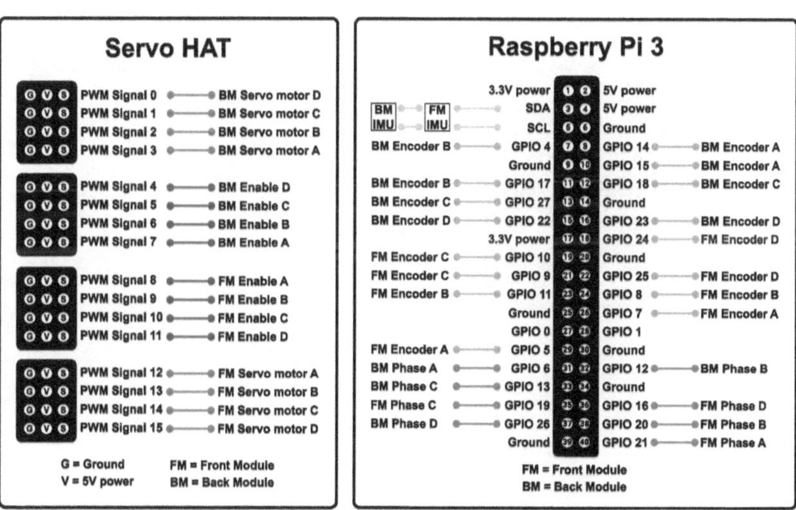

Figure 3.17: Interconnections between actuators, drivers and sensors

The red wires are digital outputs that change the rotation of the wheels, the blue wires are PWM outputs that modify the speed of the wheels, the green wires are PWM outputs that change the angle of the arms, the orange wires are digital inputs that read the encoder and know the angular position of the wheels, and the cyan wires are the I2C wires that read the orientation of the propulsive modules.

3.3.5 Control Module Structure

The dimensions of the control module are restricted by the design dimensions shown in Table 3.3, where the diameter is 76.20 mm and the distance between universal joints is 210.00 mm. Fig. 3.18 illustrates the front and the back view of the assembly, marking the places of each component. This structure has many slots along the assembly to manage all the wires between electrical components. The lids have a hole in the middle to create a 1/2-13 thread and fix the universal joints on the sides. The complete structure is divided into four parts: two lids, and two half cylinder shaped parts for easy assembling.

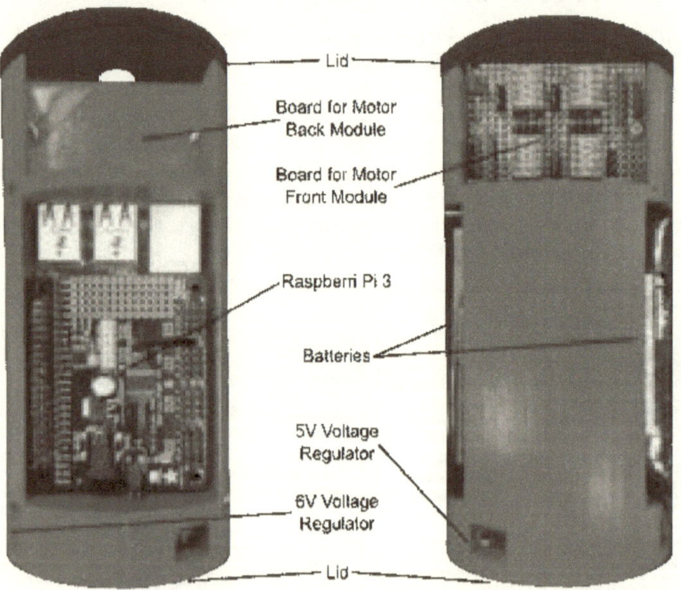

Figure 3.18: Front and back view of the control module structure

3.4 Programming

The programming was divided into five main functions called: *GUI*, *read_IMU*, *read_encoder*, *write_servo*, and *write_motor*. The *GUI* function contains the other four functions. The two actuator functions are called upon when an actuator event is selected. The two reading functions of the sensor are continuously called upon in the background. Even if an actuator event is received at the same time as a reading function, the on-board computer reads the sensors. This flow allows to have the most updated values at all times. Fig. 3.19 shows the main flow diagram of the interface.

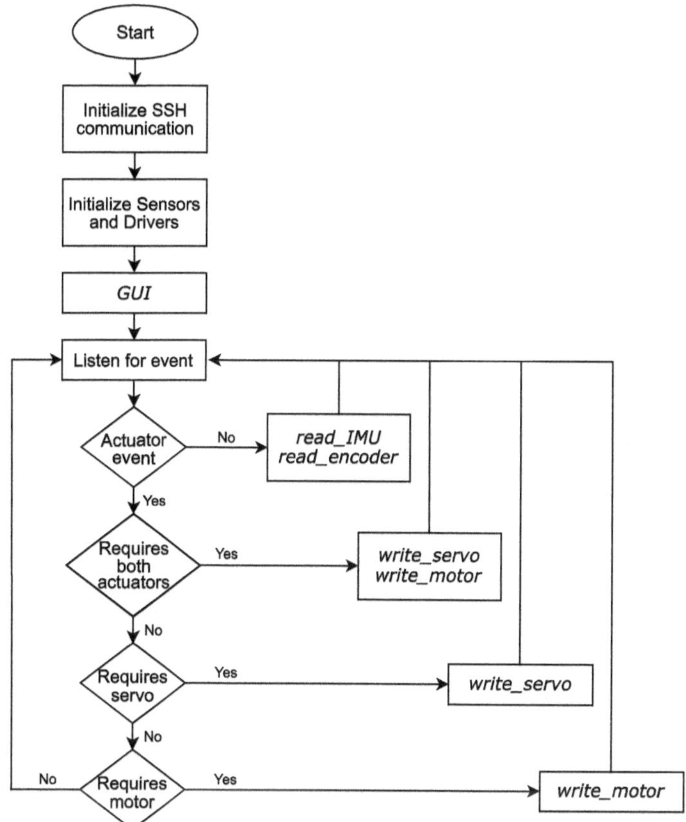

Figure 3.19: Block diagram of the GUI

The *write_servo*, and the *write_motor* functions use the Raspberry Pi outputs to control the direction of the gearmotors. They also use the Servo HAT to control the speed of the gearmotors and the position of the servomotors with a pulse width modulation (PWM). The pulse width is the time between the rising and falling edges of a pulse train.

A digital servomotor is controlled by a PWM at a frequency of 50 Hz, which means that a pulse occurs every 20 ms. The servomotor reads the PWM and moves the shaft to the correct angular position. Normally, a pulse width of 1 ms duration moves the shaft to 0°, and a pulse of 2 ms duration moves it to 180°. However, the pulse width can vary in order to reach 180°, as is the case with the EzRobot heavy-duty servomotor. After testing the servomotor with different pulse widths, it was found that 0.55 ms and 2.3 ms are the minimum and maximum pulse durations to rotate from 0° to 180°. Fig. 3.20 shows a diagram with these results.

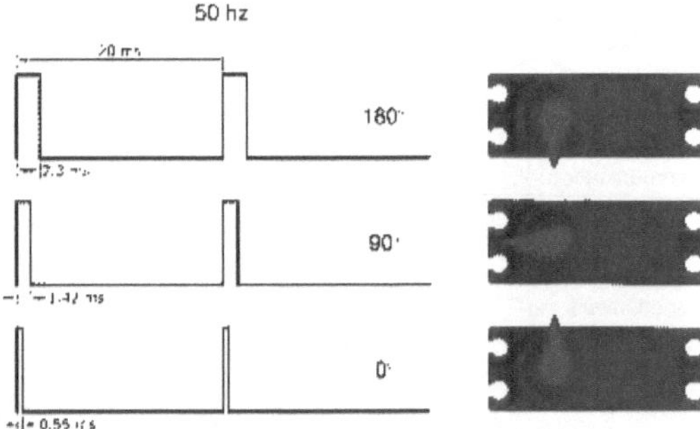

Figure 3.20: EzRobot servomotor position depending on pulse width

The gearmotor uses the same frequency of 50 Hz but the pulse width range goes from 0 ms to 20 ms, where 20 ms is equal to a full speed of 32 rpm. To change the direction of the wheels, Raspberry Pi sends 1 logical (5 V) for an anticlockwise direction and 0 logical (0 V) for a clockwise direction, to the motor drivers.

The *read_IMU* function calculates the orientation of each propulsive module. The IMU gives the angular velocity and acceleration but not the angular position. Thus, the angular acceleration readings are used to estimate the orientation with the gravity forces. To do so, these gravity measurements are converted to inclination angles, as illustrated in Fig. 3.21. This figure shows the reference position and the rotation plane with the generated components and angles used to convert the gravity measurements into inclination angles [16].

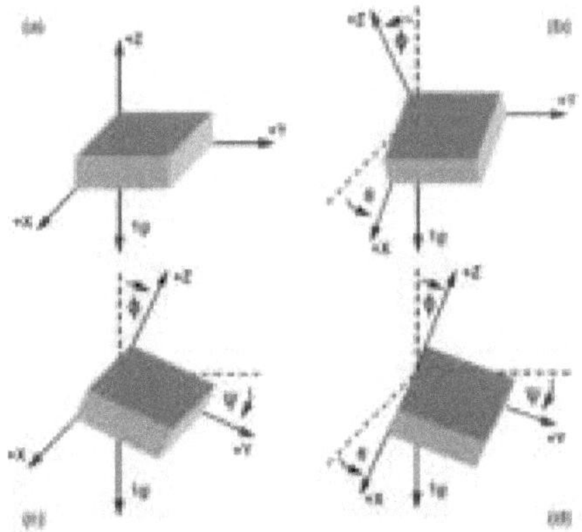

Figure 3.21: Rotation planes with components and angles [16]

The angles are calculated using:

$$\theta = \operatorname{atan2}\left(-A_{Y,OUT}, -A_{Z,OUT}\right) \tag{3.21}$$

$$\psi = \operatorname{atan2}\left(-A_{X,OUT}, -A_{Z,OUT}\right) \tag{3.22}$$

where $A_{X,OUT}$, $A_{Y,OUT}$ and $A_{Z,OUT}$ are the IMU's angular accelerations about the X, Y, and Z axis, respectively. This method assumes quasi-static or constant velocity and that the primary acceleration measurement is due to gravity.

The yield angles in Eq. 3.21 and Eq. 3.22 have an interval of $[-180°, +180°]$, where $-180°$ is the same as $+180°$. These results represent the roll and pitch angle of each propulsive module. Fig. 3.22 shows how the roll angle (ϕ) and the pitch angle (ψ) are placed in each propulsive module, as well as the direction where they increase their value.

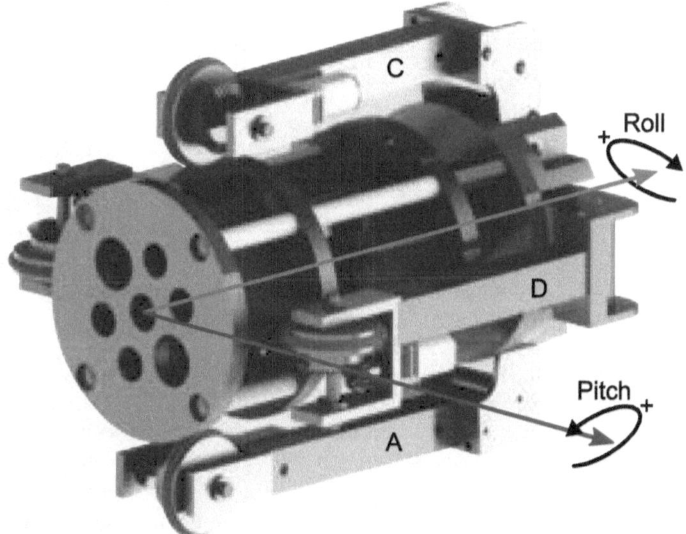

Figure 3.22: Roll and pitch angle in the propulsive module

The *read_encoder* function uses the incremental encoders to calculate the rotation of the wheels. This type of encoder has two channels, channel A and channel B. The on-board computer saves the actual state of the channels and, depending of the next channel state, the code increases or decreases the counter for that encoder. These signals are offset by 90° to differentiate between a clockwise and an anti-clockwise rotation. Fig. 3.23 shows the state diagram that was used to get the angular position of the gearmotors. There are only four states, the arrows show if the counter should add or subtract from the actual value depending on the direction of the next state. According to Section 3.3.2, when an encoder reaches 23674 counts, it means that the wheel has completed a revolution.

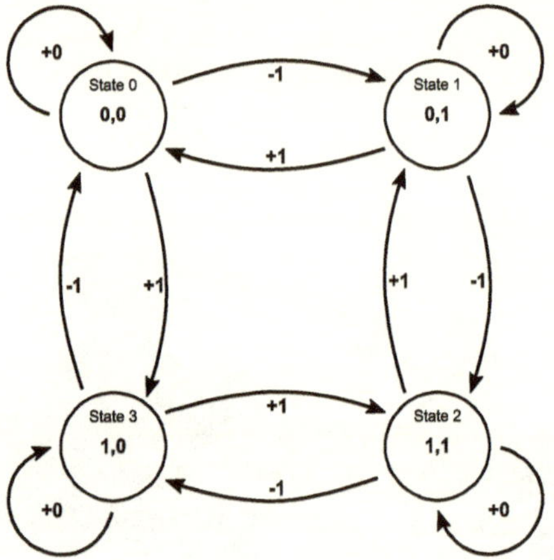

Figure 3.23: State diagram for incremental encoders

Raspberry Pi runs a script during the initialization and if the program detects a known SSID, the on-board computer automatically connects to the Wi-Fi. However, if it does not detect a known SSID, the on-board computer creates an access point for the user's connection. The data transfer in this connection is established via Secure Shell (SSH) with PuTTY.

A GUI was designed to send and receive data between the Raspberry Pi 3 B+ and the external computer. Fig. 3.24 illustrates the main view of the interface. The controls have the same labelling as used previously for wire management (Fig. 3.16). The GUI script is constantly reading the sensor values to calculate the inclination angles of the propulsive modules and the angular position of the wheels. It also sends the necessary data to the drivers when a slider bar, button or check button changes state.

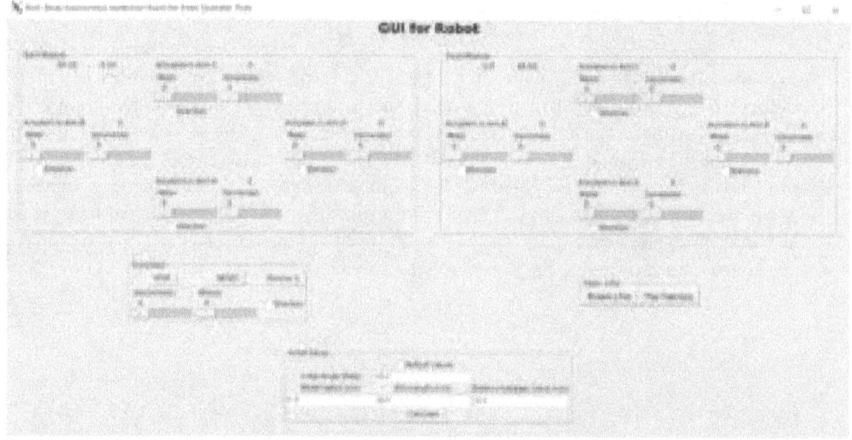

Figure 3.24: GUI to control the robot

The interface is divided in five sections and each section is represented by a square frame with a title in the top left corner. The sections at the top all have the same functionalities, expect that the left section controls the propulsive module at the back and the right section controls the propulsive module at the front. These two sections have eight slide bars, four to change the speed of the gearmotors from 0 to 32 rpm and the others to change the position of the servomotors from 0° to 180°. They also contain check buttons to individually change the rotation on the gearmotors; the robot moves forward by default.

The section in the middle-left is called "Functions" and it contains pre-programmed routines that the robot can follow, as well as general functions to control all of the servomotors and gearmotors. These functions stop the rotation of the gearmotors, reset the position of the servomotors, move the servomotors to a specific position or start the gearmotors at a specific speed.

The section in the middle-right is called "Open file" and it has three buttons. The "Browse a file" button allows the user to open the file manager and search for a text file that contains the robot's routine. The "Play Trajectory" button reads the selected text file and sends the speed or position of the actuators. Lastly, the "Stop Trajectory" button stops the execution of the selected text file.

The last section at the bottom is called "Initial Setup" and it helps the user initialize the angle of the arms for the inspection. The check button at the top is to choose the actual dimension of the robot such as wheel radius, arm length and distance between shoulder joints. In case the user is using a robot with different dimensions, these values can be inputted manually. The initial angle field is given in the text file generated by the dynamic controller. The "Calculate" button displays a pop-up message with the distance between wheels. This distance must be adjusted after the servomotors were set at 0°.

3.5   Chapter Summary

This chapter details the architecture of the proposed in-pipe robot for pipelines of different diameters. Free-body diagrams that depict the worst-case scenarios inside a pipe were developed to estimate the forces that the robot must exert. These forces, the dynamic model proposed by Douadi *et al.* [17], and the robot's dimensions were used to select the best actuators that can fit and generate the required torques. The current consumption of the components and the required voltage were utilized to select the electrical components of the robot. The electronic diagram and wiring were detailed, as well as the programming the on-board computer employs to control the robot.

Chapter 4

Simulation

In this chapter, the dimensions and the architecture of the robot, which are detailed in the previous chapter, are adapted to the kinematic model proposed by Douadi *et al.* [34]. The wheel radius ρ, and the controller are subjected to an optimization algorithm to maximize the range of pipes that the robot can inspect, and simulations are then performed.

Section 4.1 describes the controller proposed by Douadi *et al.* [34, 17]. Section 4.2 explains an optimized controller for the trajectory tracking of the robot. Section 4.3 details a series of simulations with different scenarios.

4.1 Controller

The modified kinematic model of the propulsive module is shown in Fig. 4.1. This model establishes a relation between the end-effector and the joints coordinates. The joint-coordinate vector is described by the angle of the two arms(α_r and α_l) and by the position of the two wheels' centre of mass (B_r and B_l). The end-effector coordinates have three components, the position of the module's centre of mass and the orientation of the propulsive module. Therefore, these two vectors are defined as follows:

$$x = \begin{bmatrix} x_g & y_g & \theta \end{bmatrix}^T \tag{4.1}$$

$$q = \begin{bmatrix} \alpha_r & \alpha_l & x_{B_r} & y_{B_r} & x_{B_l} & y_{B_l} \end{bmatrix}^T \tag{4.2}$$

Each propulsive module can be represented as a closed kinematic chain, expressed by the following vector relations:

$$r_{OG} = r_{OC_r} + r_{C_r B_r} + r_{B_r S_r} + r_{S_r G} \tag{4.3}$$

$$r_{OG} = r_{OC_l} + r_{C_l B_l} + r_{B_l S_l} + r_{S_l G} \tag{4.4}$$

where O is the origin, G is the centre of mass of the propulsive module, C_r and C_l are the points of contact between the wheel and the inner wall of the pipe, B_r and B_l are the centres of the wheel, and S_r and S_l are the arm joints.

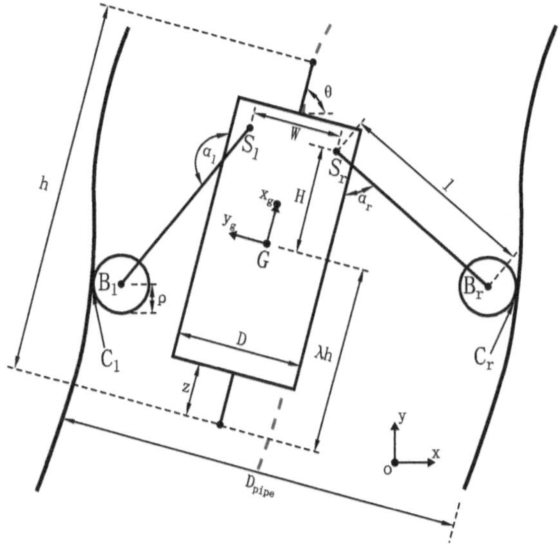

Figure 4.1: Kinematic model of a propulsive module

Douadi *et al.* [34, 17] developed a trajectory tracking controller based on the kinematics and dynamics of a multibody robot. The kinematics is described by a chain of parallel mechanisms, and the dynamics is based on Lagrange's formulation. The controller keeps track of the centreline of a pipe with the position of the centre of mass of the first propulsive module (x_{g1}, y_{g1}), and the orientation of the first and second propulsive modules (θ_1, θ_3). Therefore, the pose to control the robot is described by the following vector:

$$X = \begin{bmatrix} x_{g1} \\ y_{g1} \\ \theta_1 \\ \theta_3 \end{bmatrix} \tag{4.5}$$

The pose that the controller uses does not take into consideration the position of the centre of gravity of each module. So, the second propulsive module might not follow the centreline of the pipeline. The path to follow is calculated using the Frenet frames technique. A Frenet frame is assigned to each propulsive module by considering the projection of its centre of gravity on the centreline of the pipe. These frames are associated with the desired path composed of the unit tangent and the unit normal to the centreline. The error between the pose and the real pose is processed by a torque controller. This controller generates the required torques on the arms and on the wheels in order to track the desired path. Fig. 4.2 illustrates the controller configuration.

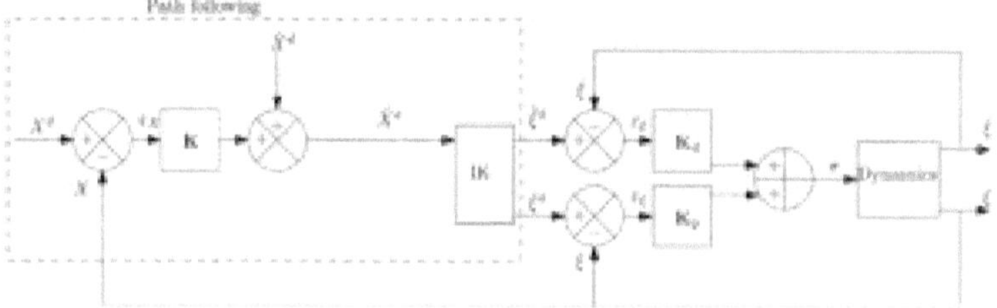

Figure 4.2: PD control scheme [17]

4.2 Optimization

The optimal values obtained by Lamonde [35] are the basis of the actual design values. However, the distance between shoulder joints is 15 % smaller than the optimal value due to the servomotor's dimensions. The change in this distance can cause an increase in the motion of the arms. To verify that the robot is able to execute the inspection in 6-inch and 8-inch pipes, an algorithm that runs the controller proposed by Douadi *et al.* [17] was developed and used in different scenarios.

This program varies the wheel radius ρ of the robot and the pipe diameter D_{pipe}, while saving the minimum and maximum aperture of the arms. Fig. 4.3 depicts the necessary range of motion that the arms must possess in order to perform such trajectories. This graph shows that the paths for 6-inch and 7-inch pipes require a range of motion larger than 19° for any wheel radius. This motion exceeds the maximum range of motion of 18° obtained in Section 3.2.2. Thus, an optimization is proposed in this section to maximize the range of pipes that the robot can inspect.

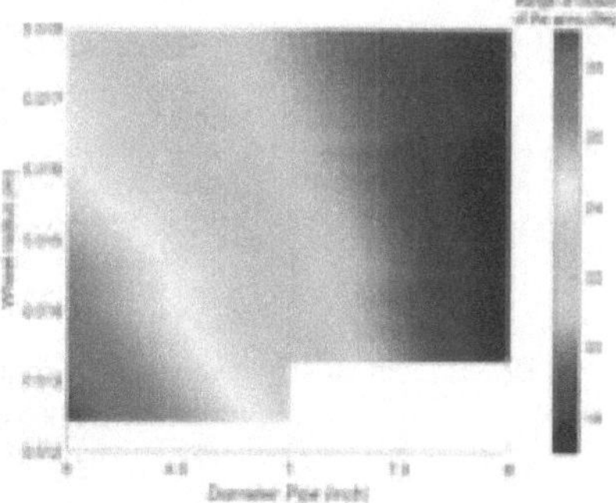

Figure 4.3: Range of motion of the arms in 6, 7, and 8 inches pipelines

The controller uses the centre of mass of the first propulsive module to track the position of the desired kinematics. This centre varies between 60 % and 65 % of the module length h, depending on the position of the arms α_r and α_l. Thus, a new controller was created in an attempt to reach the range of motion of the arms with the actual worm drive mechanism. This controller follows a new tracking point (CC) that can move along the longitudinal axis of the propulsive module instead of the centre of mass. This tracking point is defined as follows:

$$CC = \gamma\, h \tag{4.6}$$

where γ is a non-dimensional variable that can vary between 0 and 1, and h is the length of the propulsive module.

This parameter changes the position of the robot, but it does not change its orientation. Then, the new pose to control the robot is described by:

$$X = \begin{bmatrix} x_{CC} \\ y_{CC} \\ \theta_1 \\ \theta_3 \end{bmatrix} \tag{4.7}$$

Hence, a new algorithm that iterates the value of gamma γ, the wheel radius ρ, and the pipe diameter D_{pipe} was developed. The aim of this program is to find the best values of ρ and γ to inspect pipes of 6-inch to 8-inch of diameter. This algorithm uses a 90° short radius elbow because this configuration is the most difficult configuration for the robot to travel. Table 4.1 shows the range of parameters that the algorithm uses.

Table 4.1: Range of the parameters for optimization algorithm

Parameter	Minimum Value	Maximum Value	Increment
γ	0.30	0.70	0.05
ρ	0.010 m	0.018 m	0.001 m
D_{pipe}	6-inch	8-inch	1-inch

At the end of each pipe simulation, the program compares the range of motion of the different scenarios and displays a graph with the results. Fig. 4.4 depicts the block diagram of the algorithm.

The graphs produced by the algorithm show the range of motion of the arms, the wheel radius ρ, and the value of γ. Fig. 4.5 depicts the results of the program in a 6-inch pipe. In this scenario, the minimum range of motion required by the arms is 21°, which is smaller than the results from the original controller, however the maximum arm motion that the servomotors can exert is 18°. Thus, it confirms that the actual mechanism with the actual controller is not able to navigate through 6-inch short radius fittings.

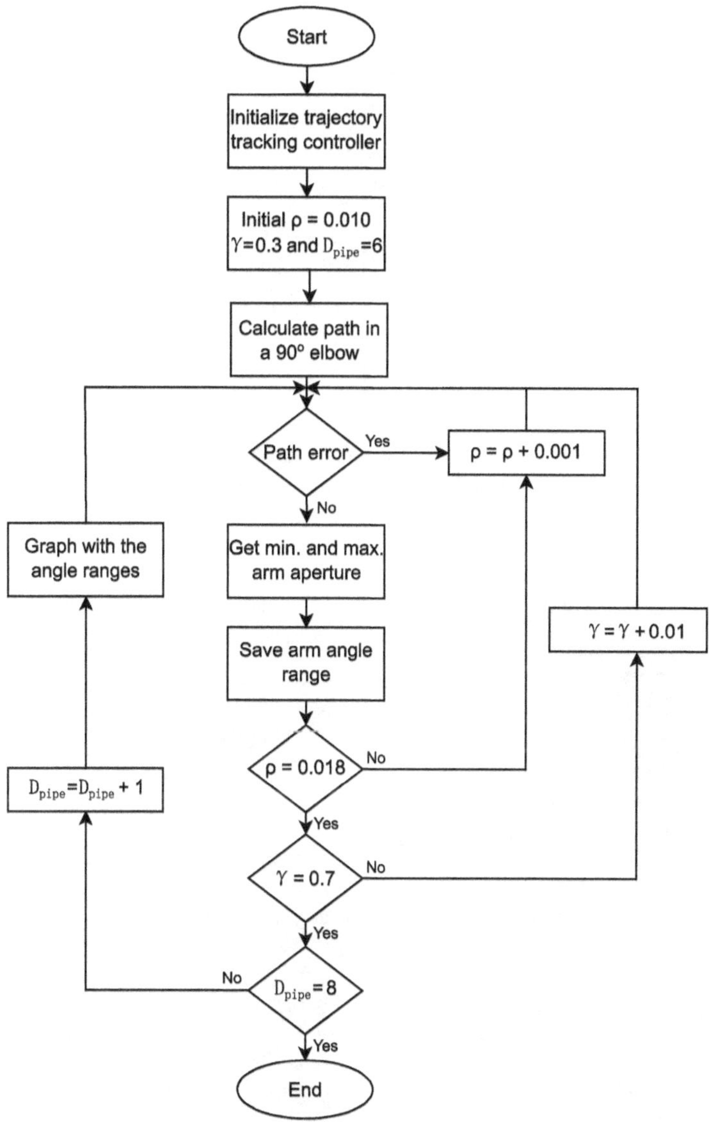

Figure 4.4: Block diagram to optimize the wheel radius ρ, and the value of γ

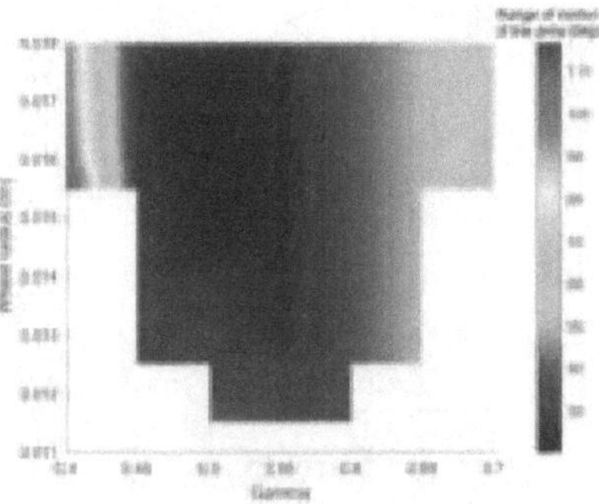

Figure 4.5: Range of motion of the arms in 6-inch pipes

Fig. 4.6 illustrates the results of the simulation in a 7-inch pipe and Fig. 4.7 shows the results for an 8-inch pipe. These two graphs demonstrate that the robot can inspect standard pipes between 7 and 8 inches of diameter because the range of motion of the arms is within the limits of the servomotors.

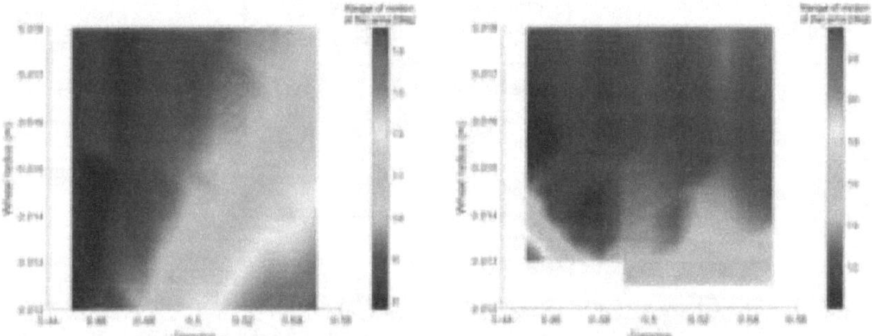

Figure 4.6: Range of motion of the arms in 7-inch pipes

Figure 4.7: Range of motion of the arms in 8-inch pipes

Based on the results from inspections in 6-inch and 7-inch pipes, the best value for gamma must be between $\gamma = 0.45$ and $\gamma = 0.50$. Conversely, the results from the inspections in 8-inch pipes show that the best value for the wheel radius must be between ρ=0.015 m and ρ=0.016 m. Therefore, the optimal values for the new controller are set at γ=0.45 and ρ=0.0155 m.

4.3　Robot Navigation

In this section, three simulations in 8-inch pipelines are presented to analyze if the selected actuators can execute such trajectories. The first simulation is in a reduction pipe to 7 inches, the second one is in a 90° short radius elbow and the third one is in a 180° short radius elbow.

The simulations were programmed in Matlab in a two-dimensional environment because it is assumed that the robot does not rotate inside the pipe. Therefore, the presented outputs are for the left and right arms of the propulsive modules. The pipe dimensions are based on standard pipes (See Appendix A) without defects in their inner walls.

The controller has three important outputs, the torques on the wheels and the arms, the linear velocity on the centre of each wheel and the angle of each arm. The linear velocity and the angle of the arms are modified with the angular velocity of the gearmotors and the position of the servomotors. Thus, these variables were merged into a single graph with two different scales in the y-axis.

Each graph serves a specific purpose in the verification of the selected actuators and of the design of the robot. The position of the arms cannot exceed the mechanical range of 11° to 43°. The position of the servomotors must be smaller than 180°, which is the maximum motion of these actuators. The rotational speed of the gearmotors may not surpass the 32 rpm that the selected motor can provide. Lastly, the torques on the wheels and on the arms shall not pass the maximum torque performed by the actuators, which is 1.5 Nm and 6.75 Nm, respectively.

4.3.1 8-inch to 7-inch pipe reduction

Fig. 4.8 depicts five important robot positions inside the pipeline during the transition from an 8-inch to a 7-inch pipe. In this scenario, the robot does not need to change its orientation to track the path, but it must adjust the angle of the arms to fit into the new pipe diameter. Fig. 4.9 depicts the distance error of both modules with respect to the pathline. Since the robot does not changes its orientation, the position error in both modules is minuscule.

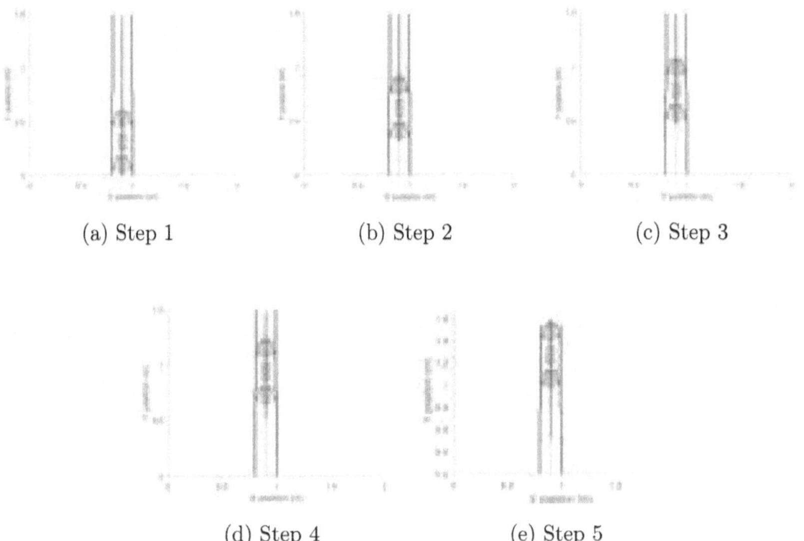

(a) Step 1   (b) Step 2   (c) Step 3

(d) Step 4   (e) Step 5

Figure 4.8: Steps to travel inside a reduction from 8 to 7 inches

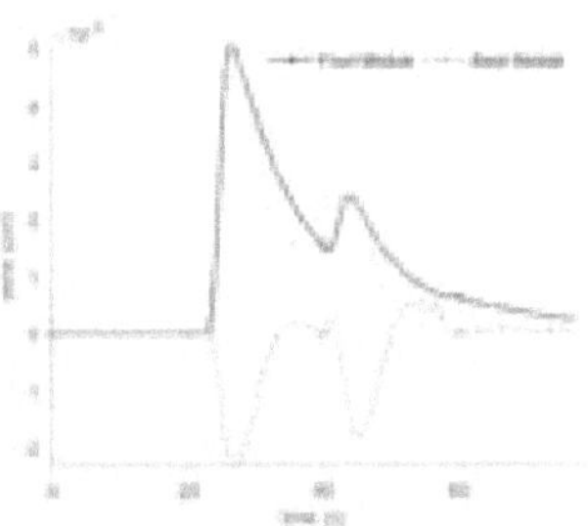

Figure 4.9: Distance errors inside a reduction from 8 to 7 inches

Fig. 4.10 shows the required torques during the inspection. The maximum torque is applied in the arms when the robot is adjusting the arms into the new diameter and it has a value of 0.15 N m.

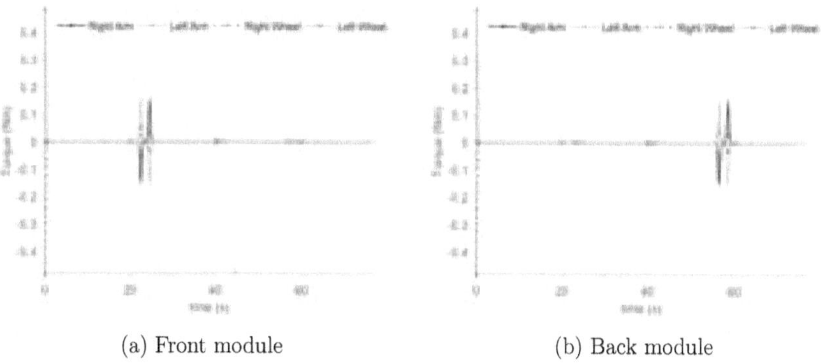

(a) Front module

(b) Back module

Figure 4.10: Torques of the arms and of the wheels in a reduction from 8 to 7 inches

Fig. 4.11 shows the gearmotors speed and the linear velocity on the centre of the wheels. These graphs show that the robot goes at a maximum inspection speed of 2.5 cm/s until the propulsive module faces the reduction, at which point the speed decreases to 2.1 cm/s. During this inspection the robot did not exceed the 32 rpm limits of the gearmotor.

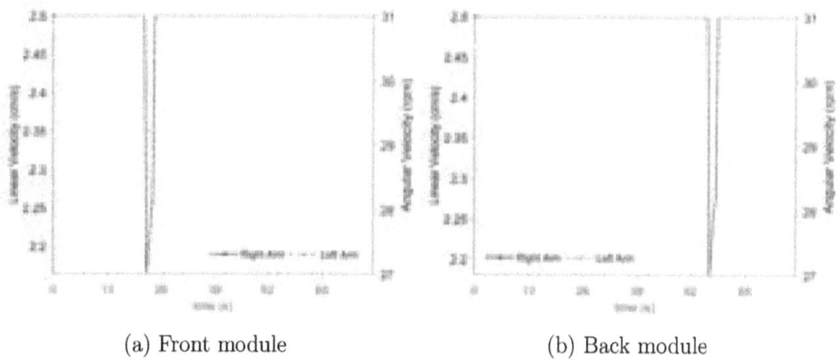

(a) Front module

(b) Back module

Figure 4.11: Angular velocity of the gearmotors and linear velocity of the wheels in a reduction from 8 to 7 inches

Fig. 4.12 shows the angular position of the servomotors and of the arms. The motion of the arms goes from 32° to 25°, while the servomotor goes from 64° to 2°, which is below the range of motion of the servomotors.

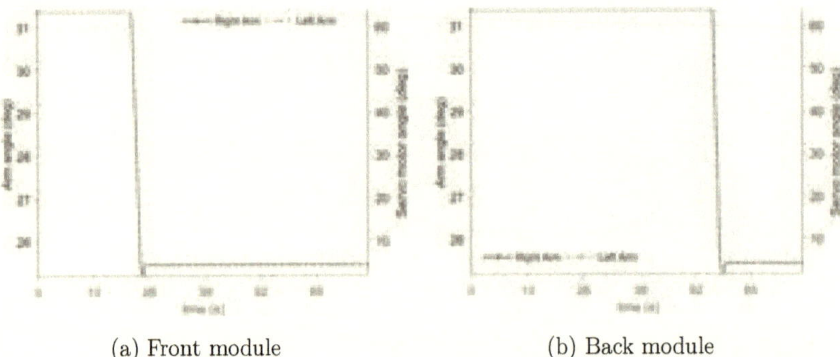

(a) Front module   (b) Back module

Figure 4.12: Position of the arms and servomotors in a reduction from 8 to 7 inches

The inspection in a reduction pipe shows that the front and back propulsive modules perform the same or very similar movements. This situation is also seen in the previous graphs for the front and back modules, where the torque, velocity, and position are similar in magnitude but are executed at different times, as the front propulsive module enters first and the back propulsive module enters later.

4.3.2   8-inch elbow of 90°

Fig. 4.13 shows five of the important positions for the robot as it travels through an 8-inch short radius elbow of 90°. The controller keeps the centre of the first propulsive module on the centreline almost all of the time. However, sometimes the second propulsive module cannot reach the path. This situation is observed in Fig. 4.14, where the back propulsive module reached a position error of almost 2.7 cm with respect to the path.

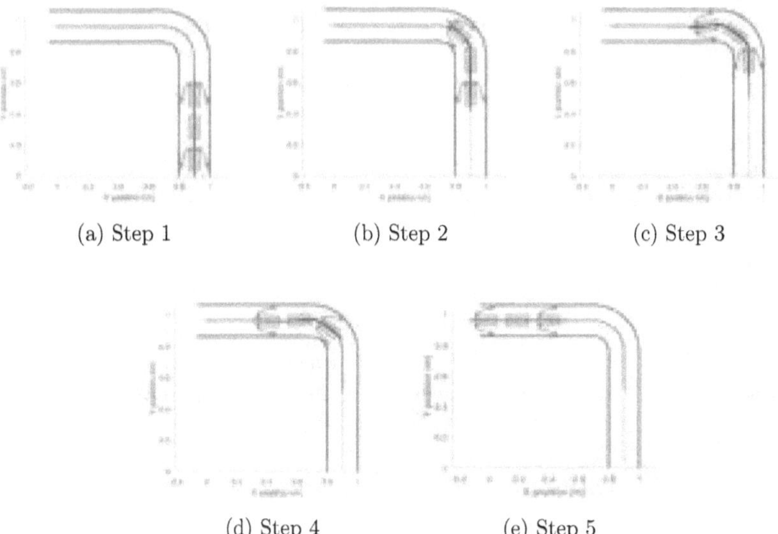

(a) Step 1   (b) Step 2   (c) Step 3

(d) Step 4   (e) Step 5

Figure 4.13: Steps to travel inside an 8-inch short radius elbow of 90°

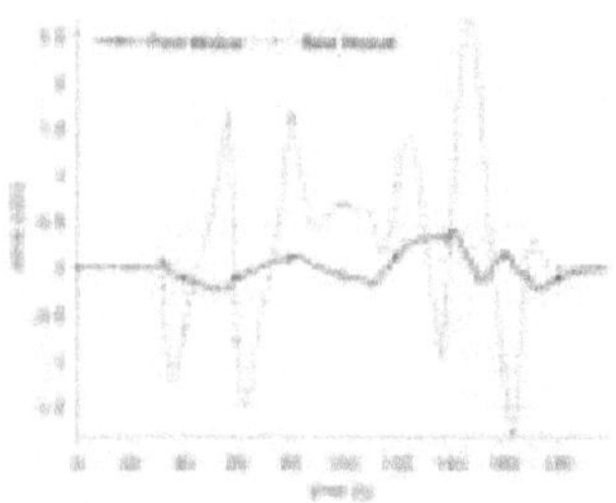

Figure 4.14: Distance errors inside an 8-inch short radius elbow of 90°

Fig. 4.15 depicts the torques on the front and the back modules during the inspection in this pipeline. The maximum torque on the wheels is of 0.32 N m, while the maximum torque on the arms is of 0.24 N m.

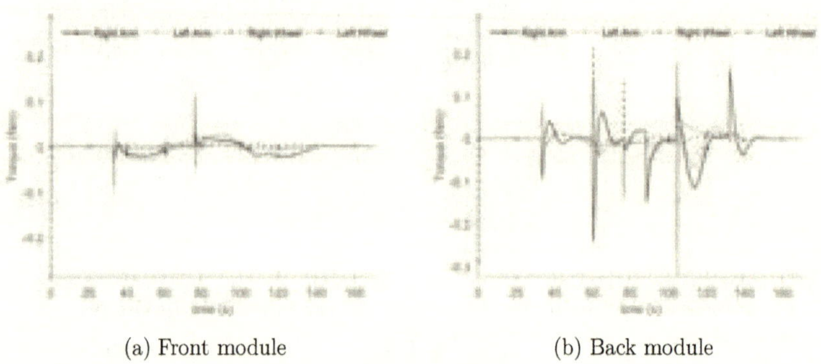

(a) Front module

(b) Back module

Figure 4.15: Torques on the arms and the wheels in an 8-inch elbow of 90°

Fig. 4.16 shows the gearmotors speed and the linear velocity on the centre of the wheels. The inspection speed in this simulation is of 1.5 cm/s, but the maximum linear speed on the wheels reaches 2.3 cm/s, which is equivalent to 29 rpm for the gearmotors.

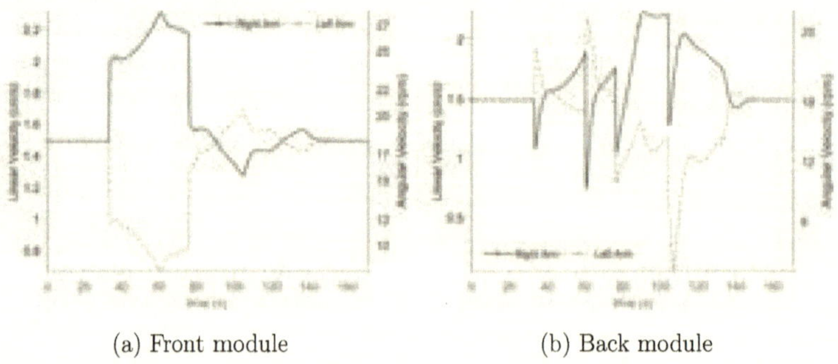

(a) Front module

(b) Back module

Figure 4.16: Angular velocity of the gearmotors and linear velocity on the wheels in an 8-inch elbow of 90°

Fig. 4.17 illustrates the angular position of the servomotors and the arms in each propulsive module. The minimum angle of the arms is 25° and the maximum angle is 39°, which represents a range of motion of 14°.

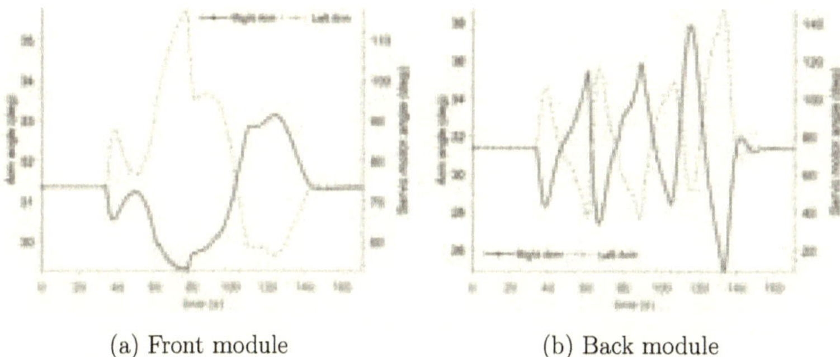

(a) Front module   (b) Back module

Figure 4.17: Position of the arms and servomotors in an 8-inch elbow of 90°

The previous figures show that the actuators in the back propulsive module generated larger torques, velocities and movements than the first propulsive module. This difference is caused by the kinematics constraints and the proposed controller. However, the selected actuators are able to manage the required inputs in order to complete the navigation through an 8-inch elbow pipeline with a short radius.

4.3.3   8-inch elbow of 180°

Fig. 4.18 shows five of the important positions during an inspection in an 8-inch elbow of 180° with a short radius. The behaviour of the first and second propulsive modules is similar to the behaviour of the modules in the previous simulation. Fig. 4.19 depicts the position error of this simulation, where the robot reached an error of 2.5 cm with respect to the trajectory.

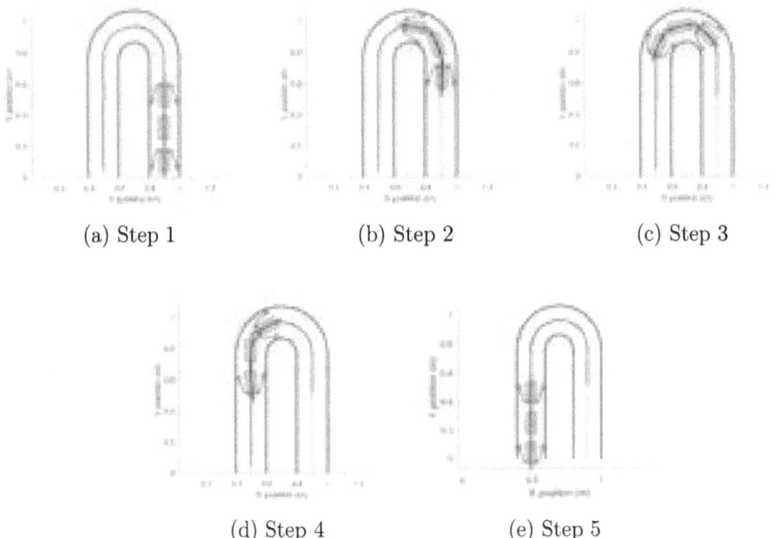

<table>
<tr><td>(a) Step 1</td><td>(b) Step 2</td><td>(c) Step 3</td></tr>
<tr><td>(d) Step 4</td><td>(e) Step 5</td><td></td></tr>
</table>

Figure 4.18: Steps to travel inside an 8-inch elbow of 180° with a short radius

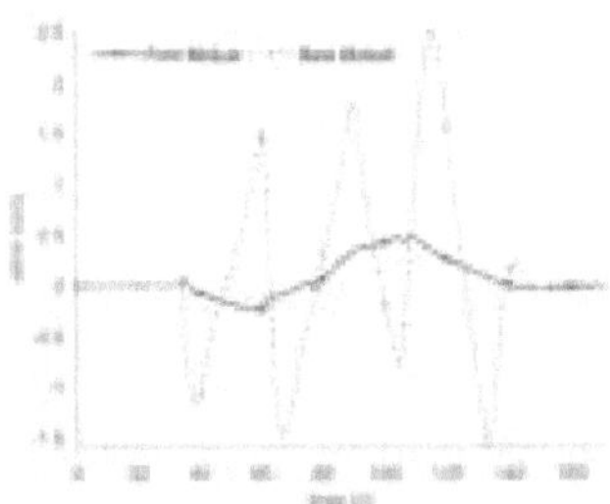

Figure 4.19: Distance errors inside an 8-inch short radius elbow of 180°

Fig. 4.20 shows the torques on the front and back modules. The maximum torque on the wheels is of 0.35 N m and the maximum torque on the arms is of 0.26 N m.

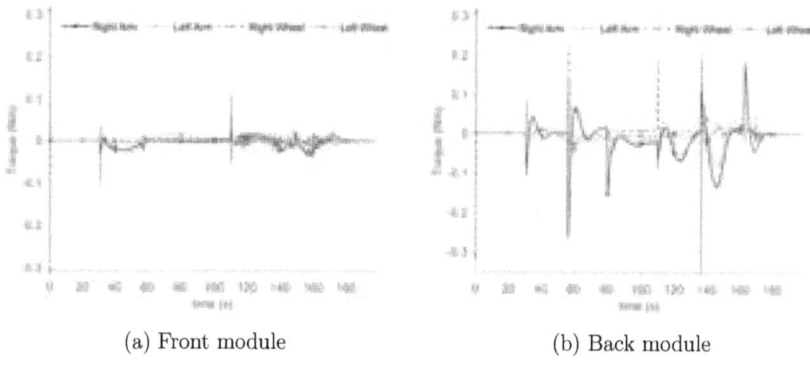

(a) Front module   (b) Back module

Figure 4.20: Torques on the arms and the wheels in an 8-inch elbow of 180°

Fig. 4.21 depicts the gearmotors speed and the linear velocity on the centre of the wheels in each propulsive module. The inspection speed of the robot in this simulation reaches 1.6 cm/s, while the maximum speed of the gearmotors is of 30.5 rpm.

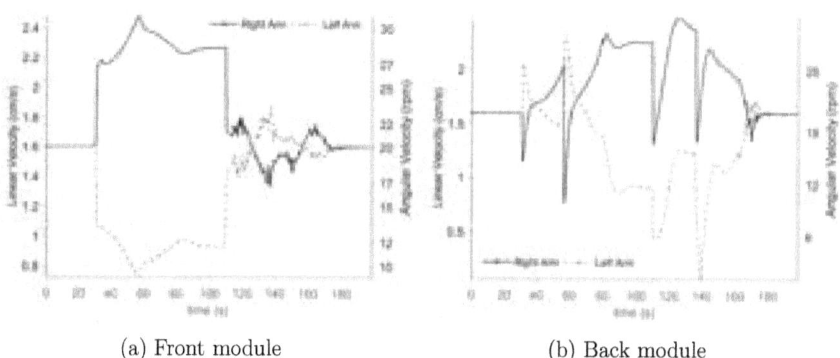

(a) Front module   (b) Back module

Figure 4.21: Angular velocity of the gearmotors and linear velocity on the wheels in an 8-inch elbow of 180°

Fig. 4.22 shows that the minimum and maximum angles of the arms are 24° and 40°, respectively, which means that the robot's arms require a range of motion of 16°. The maximum angle of the arms is close to the maximum mechanical angle but it does not exceed it. The servomotor position varies from 0° to 154°, which is the largest motion in all of the simulations, but it is still in the capabilities of the servomotor.

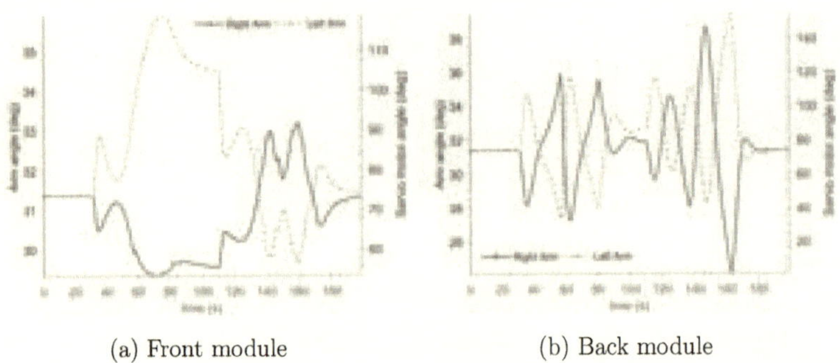

| (a) Front module | (b) Back module |

Figure 4.22: Position of the arms and servomotors in an 8-inch elbow of 180°

4.4 Chapter Summary

In this chapter, the radius of the robot and gamma, the non-dimensional variable that modifies the tracking point of the controller, are optimized to enhance the performance of the robot in various pipe diameters. Furthermore, a series of simulations using the optimized variables were developed. The minimum and maximum positions of the servomotor are 0° and 154°, respectively. It is also shown that the maximum speed required by the gearmotors is of 30.5 rpm. These values are within the capabilities of the actuators, where the total motion of the servomotor is of 180° and the maximum speed of the gearmotor is of 32 rpm. Therefore, the robot is able to overcome any type of single 7-inch to 8-inch diameter elbow, as long as the pipe dimensions are standard. Table 4.2 depicts the most important results of the simulations.

Table 4.2: Simulations results

Pipe type	Arm Angle (deg)		Servo Angle (deg)		Max. Torques (Nm)		Motor Speed (rpm)	Robot Speed (cm/s)
	Min.	Max.	Min.	Max.	Arm	Wheel	Max.	Average
Reduction	25	32	2	64	0.15	0.08	31	2.5
Elbow of 90°	25	39	8	147	0.24	0.32	29	1.5
Elbow of 180°	24	40	0	154	0.26	0.35	30.5	1.6

Chapter 5

Experimental Tests

This chapter presents the physical prototype along with the experimental process that was used to test the robot. It also details a series of experiments where the robot was subjected to the generated outputs of the optimized dynamic controller and the resulting paths are compared to the simulated paths. For the first three open-loop experiments, the robot executed trajectories inside and outside the pipe to measure the precision of the encoders and actuators. A closed-loop experiment was also implemented to measure the effectiveness of the robot using a PD controller without the dynamic model. The last experiment estimates the autonomy of the robot when it is travelling inside the pipeline.

5.1 Experimental Platform

For the open-loop experiments, the external computer sends a text file that contains the signals of the actuators. Once the on-board computer receives the order to start the motion, it translates the file instructions into commands for the actuators. For the closed-loop experiment, the on-board computer reads the encoders and estimates the error of the robot. This error is used by the PD controller to send the required speed to the gearmotors. For the autonomy experiments, the robot executed a modified open-loop test inside the pipe, while the elapsed time was measured. Fig. 5.1 shows the assembled robot that was proposed in Section 3.1.

Figure 5.1: Assembled robot

The robot performed the experiments in a galvanized duct pipe of 8 inches. The internal dimensions of the straight pipe differ by 2 mm from the standard pipelines. Nevertheless, this difference is within the limits of the actuators of the robot. The dimensions of the pipe limit the possible tracking systems that can be used because there is not enough space in the pipe to add external sensors that can measure the position of the robot. Thus, the sensors in the propulsive modules were used as a first approach. To evaluate wheel slippage, a plastic ruler was placed inside the pipe to measure the real travelled distance in conjunction with the in-built gearmotor encoders.

5.1.1　Data Transmission

The on-board computer receives and sends information to the external computer through the Wi-Fi connection. Using this connection, a text file is sent by the external computer and is saved in the on-board computer to avoid transmission delays between commands to the actuators. This file contains the angle of the servomotors in degrees and the speed of the gear motors in rpm. The on-board computer transforms these values into inputs for the motor drivers to perform the necessary motions for the robot.

Each encoder uses an independent script that reads an encoder every time that it detects a change in the channels, while the IMUs send a new value every half second after the GUI is open. These sensors create a new text file with two values in every row; the first value is the timestamp of when the reading was executed and the second value is the sensor reading. When the locomotion is done, the user can send the generated files to the external computer in order to process the information in Matlab. The whole process, from the creation of the text file to the processing of information, is shown in Fig. 5.2.

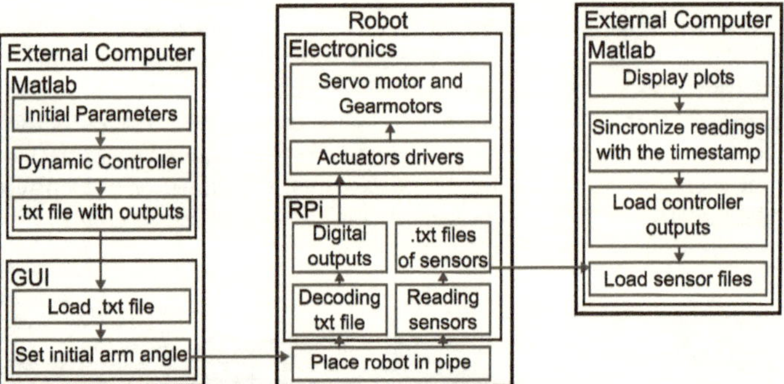

Figure 5.2: Experimental process

5.2 Robot Experiments

The prototype was subjected to multiple paths to analyze the difference between the expected trajectory, the measured trajectory and the real trajectory. There are two encoder tests outside the pipe, the first executed a linear motion, while the second executed an angular motion. The third and fourth tests are open-loop and closed-loop motion tests inside an 8-inch straight pipe. Finally, the last test is an endurance test to estimate the autonomy of the robot.

5.2.1 Encoder Test in Linear Motion

This experiment analyzes the precision of the encoders when the gearmotors rotate at a constant speed. The on-board computer reads the file created in Matlab, containing the speed in rpm, and it decodes the speed into PWM signals for the drivers. This file contains four values, two for the right arms of the robot and two for the left arms. For this test, the robot is placed outside the pipe in an orientation where the wheels do not have any contact with a surface, in order to avoid propulsion. Thus, the only load for the gearmotors is the weight of wheels.

The initial and final conditions for this test are shown in Table 5.1. The roll and pitch angles are considered "free" because the robot is not moving inside the pipe. Thus, the orientation of the propulsive modules is not considered important for this test. The initial conditions and the physical parameters of the robot are the inputs of the optimized controller presented in Section 4.2.

Table 5.1: Simulation parameters for the encoder test in linear motion

Time	Revolutions (rev)	Rotational Speed (rpm)	ϕ (deg)	ψ (deg)
Initial	0	15.5	free	free
Final	10	15.5	free	free

In order to measure the angular position of each wheel and send the signals to the actuators, a parallel sub-process that reads the encoders was executed by the on-board computer. The experiment was repeated ten times and the tracked positions were plotted alongside the simulation path. The plots were divided into the right and left wheels of the propulsive modules because in a 2D environment the wheels on each side of the robot rotate the same revolutions at the end of the path. Fig. 5.3 shows the revolutions of the wheels on the front module versus the revolutions of the wheels on the back module. These figures have markers at 50 % and at 100 % of the experiment duration time.

On average, at the halfway point of the simulation, the error is 4.8 % fewer revolutions on the gearmotors. While, at the end of the experiment, the error is 5.4 % fewer revolutions. Table 5.2 presents the main results of the experiment.

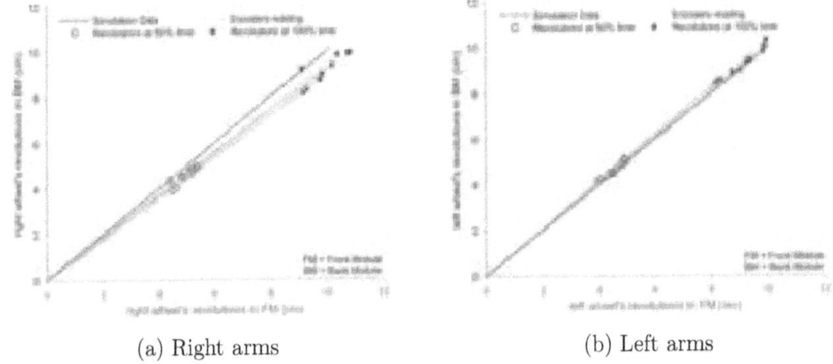

<table>
<tr><td>(a) Right arms</td><td>(b) Left arms</td></tr>
</table>

Figure 5.3: Simulated and tracked trajectories for the encoder test in a linear motion

Table 5.2: Experimental results for encoder test in a linear motion

	50 % Time		100 % Time	
	Revolutions (rev)	Rotational Speed (rpm)	Rotational Distance (rev)	Rotational Speed (rpm)
Simulated state	5	15.5	10	15.5
Average error right wheels	-0.18	-0.55	-0.41	-0.63
Average error left wheels	-0.31	-0.96	-0.68	-1.05
Average error wheels	-0.24	-0.75	-0.54	-0.84

Normally, the output and efficiency of every motor are different, even for the same models [39]. In this case, it is clear that the motors rotate at a slightly different speed when the on-board computer sends them the same PWM signal. Fig. 5.3 (b) shows that the left wheels rotate almost at the same speed, while Fig. 5.3 (a) demonstrates that the right wheel on the back module rotates at a higher rate than the right wheel on the front module. In both cases, the percentage of error, at the middle and at the end of the simulation, did not increase proportionally with respect to time. This means that the error is likely due to small differences between the motor's real performance and the motor's specifications. However, the dynamic model of the robot can include the gearmotor model in order to decrease the error between the simulation and the real trajectories.

5.2.2 Encoder Test in Angular Motion

This experiment analyzes the precision of the encoders, when the robot simulates a movement through a 90° elbow. This motion requires that the speed of the gearmotors change during the execution, which means that they cannot rotate at a constant speed.

The initial position of the robot is the same as in the previous experiment. The start and end parameters for this test are shown in Table 5.3. Once again, the roll and pitch angles are considered "free", as the robot is not moving. The initial conditions in this experiment were divided into right and left wheels because the distance travelled by the right and left wheels is different. In addition, it should be noted that this experiment uses linear distance as a parameter instead of revolutions.

Table 5.3: Simulation parameters for the encoder test in an angular motion

	Time	Linear Distance (cm)	Rotational Speed (rpm)	ϕ (deg)	ψ (deg)
Right wheels	Initial	0	9.2	free	free
	Final	274.57	9.2	free	free
Left wheels	Initial	0	9.2	free	free
	Final	222.00	9.2	free	free

Fig. 5.4 (a) shows the speed of the gearmotors that is sent by the on-board computer when it reads the text file created in Matlab. Fig. 5.4 (b) depicts the travelled distance of the right and left wheels when executing the created path.

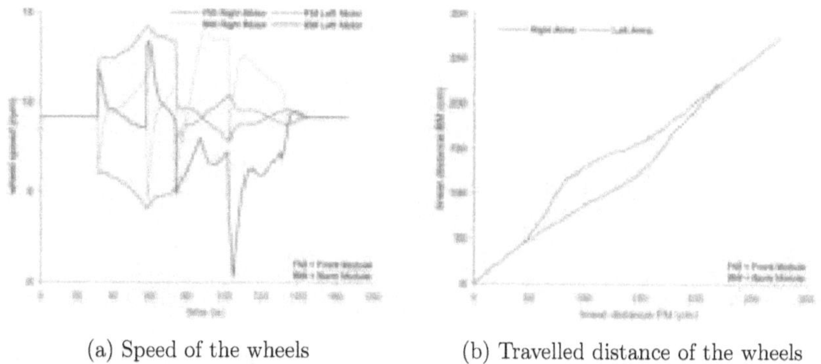

(a) Speed of the wheels   (b) Travelled distance of the wheels

Figure 5.4: Simulation of a locomotion inside an 8-inch elbow of 90°

The angular position of the wheels was measured by the encoders and saved by the on-board computer. This experiment was repeated ten times outside the pipe and the tracked positions were plotted alongside the simulated travelled distance of the wheels. Thus, the plots were divided into right and left wheels. Fig. 5.5 illustrates the wheels' linear displacement for the front module versus the wheels' linear displacement for the back module. For an easy evaluation, the plots use markers at 50 % and at 100 % of the experiment duration time.

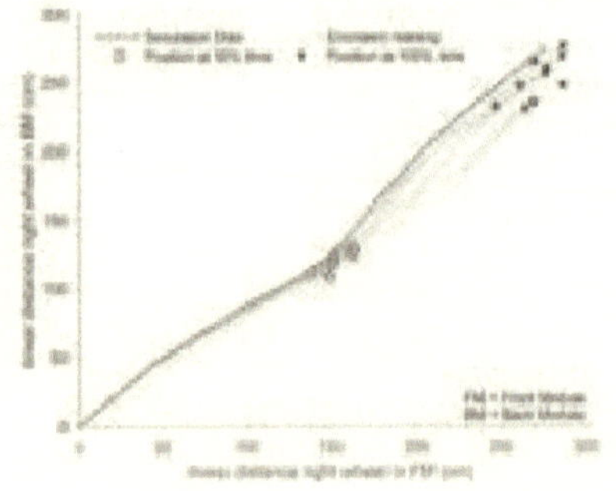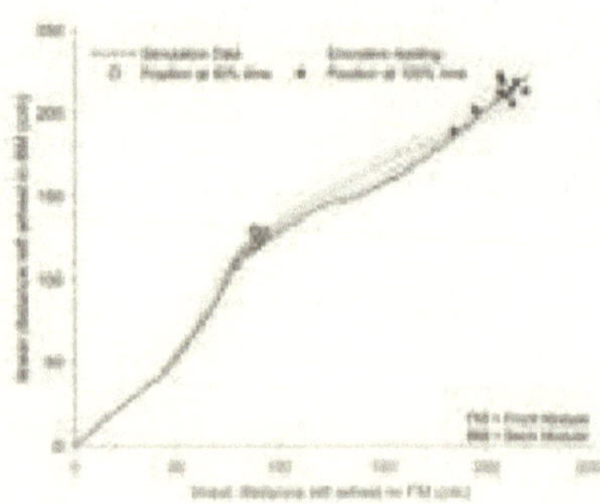

(a) Right encoders of the robot   (b) Left encoders of the robot

Figure 5.5: Simulated and tracked trajectories for the encoder test in an angular motion

On average, at the halfway point of the simulation, after 83.5 seconds, the error is 2.55 % fewer revolutions of the gearmotors. While, at the end of the experiment, after 167 seconds, the error is 4.68 % fewer revolutions. Table 5.4 presents the analytical results of the experiment.

Table 5.4: Experimental results for the encoder test in an angular motion

	50 % Time		100 % Time	
	Linear Distance (cm)	Rotational Speed (rpm)	Linear Distance (cm)	Rotational Speed (rpm)
Simulated state right wheels	139.14	9.2	274.57	9.2
Average error right wheels	-3.02	-0.19	-12.47	-0.41
Simulated state left wheels	108.51	9.2	222.01	9.2
Average error left wheels	-4.08	-0.34	-13.21	-0.54
Average error wheels	-3.55	-0.26	-12.85	-0.47

In this experiment, the error at the halfway point of the test is less than 3 %, which is considered acceptable for any type of motion. However, the error at the end is almost double that of the midpoint. It can be seen in Fig. 5.5 (a) and (b) that the error starts growing when the speed of the gearmotors is not constant. Thus, the increase in the error is caused by the change in the speed. The wheels in this experiment must vary their speed from 1 to 15 rpm to perform the required motion. This speed range can only be obtained when the gearmotors adjust their speed from 6.45 % to 96 % of their maximum speed.

This test showed speed and torque limitations when using an open-loop controller with PWM as input for the gearmotors because the selected motors do not proportionally rotate according to the input signal. Therefore, the robot needs a closed-loop controller that uses the in-built gearmotor encoders to adjust the speed of the motor.

5.2.3 Open-Loop Motion Test in a Straight Pipe

This motion test inside an 8-inch pipeline has two main purposes, the first is to evaluate the precision of the encoders and gearmotors when the robot moves in open-loop mode inside a straight pipe, and the second is to analyze the orientation of the robot when it executes the motion without the feedback of the sensors.

The initial position for this test is shown in Fig. 5.6. For this position, the robot is immobilized with one arm at the bottom of the propulsive module and the other three arms holding the robot in the centreline of the pipe. The robot must move along the pipe for 100 cm. The initial and final conditions are shown in Table 5.1.

In this experiment, the roll and pitch angle are considered important. The roll angle is used to analyze if the propulsive modules twist inside the pipe during the locomotion. While, the pitch angle is used to evaluate how the inexistence of a 3D dynamic model will affect the vertical stability of the robot.

Figure 5.6: Initial position of the robot for the open-loop motion test

Table 5.5: Simulation parameters for the open-loop motion test

	Time	Linear Distance (cm)	Rotational Speed (rpm)	ϕ (deg)	ψ (deg)
Right wheels	Initial	0	15.5	90	90
	Final	100	15.5	90	90
Left wheels	Initial	0	15.5	90	90
	Final	100	15.5	90	90

The gearmotors and servomotor signals were sent to the robot to perform a locomotion in a straight pipe for 100 cm. The on-board computer saved the angular position of the wheels and the orientation of the propulsive modules with a parallel sub-process that ran at the same time as the GUI. The experiment was repeated ten times and the tracked positions were plotted alongside the simulation path. Since this motion was conducted inside the pipeline, the ruler placed in the pipe was used to measure the real travelled distance. Fig. 5.7 shows the wheels' linear displacement for the front module versus the wheels' linear displacement for the back module. These figures have markers to identify the encoder readings at 50 % and at 100 % of the time. Moreover, they have another marker to show the travelled distance measured by the ruler.

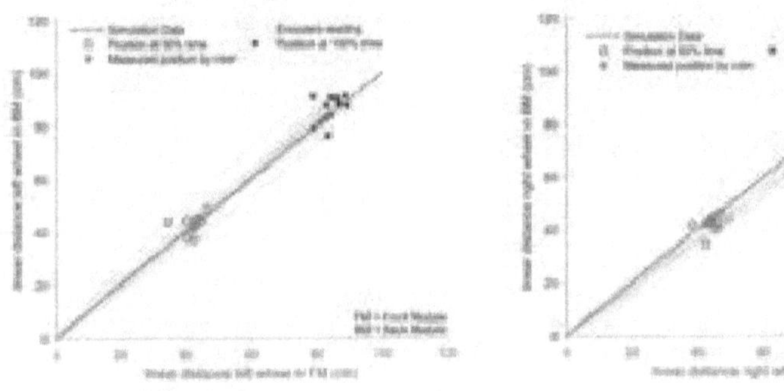

(a) Right encoders of the robot   (b) Left encoders of the robot

Figure 5.7: Simulated and tracked trajectories of the encoders for the open-loop motion test

At the halfway point of the test, after 20 seconds, the encoder error is 14.04 % less linear distance. At the end of the experiment, after 40 seconds, the error is 13.43 % less linear distance. However, at the end of the test the real travelled distance measured by the ruler is 17.00 % less than in the simulation. Table 5.6 presents the analytical results of the linear displacement in this experiment.

Table 5.6: Results of the encoders for the open-loop motion test

	50 % time		100 % time			
	Encoders		Encoders		Ruler	
	Linear	Rotational	Linear	Rotational	Linear	Robot
	Distance (cm)	Speed (rpm)	Distance (cm)	Speed (rpm)	Distance (cm)	Speed (cm/s)
Simulated state	50	15.5	100	15.5	100	2.5
Average encoder error right wheels	-6.72	-2.08	-13.48	-2.08	-	-
Average encoder error left wheels	-7.33	-2.27	-13.39	-2.07	-	-
Average encoder error wheels	-7.02	-2.17	-13.43	-2.07	-	2.16
Average ruler error	-	-	-	-	17.8	2.06

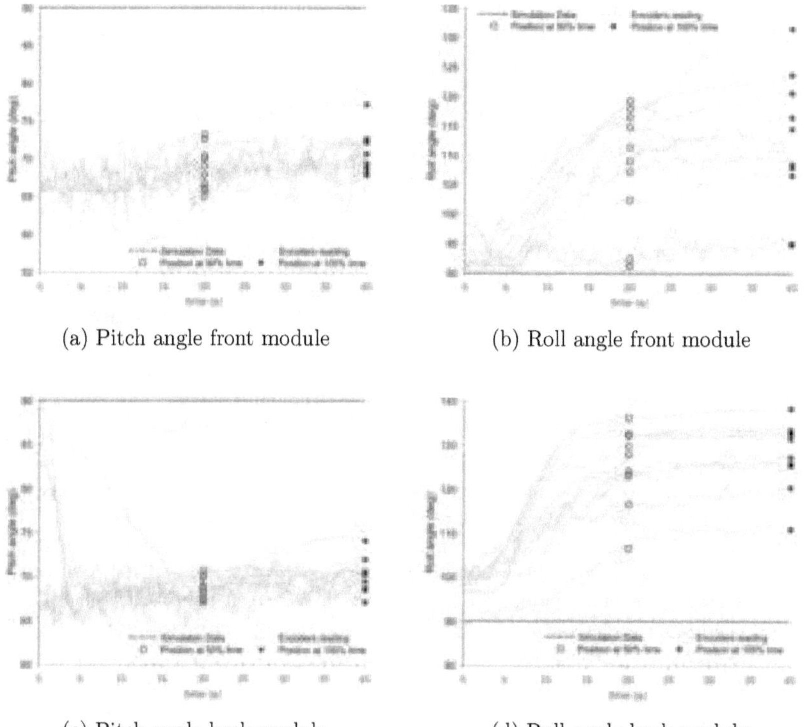

(a) Pitch angle front module   (b) Roll angle front module

(c) Pitch angle back module   (d) Roll angle back module

Figure 5.8: Simulated and tracked angles as a function of time for the open-loop motion
test

Fig. 5.8 depicts the orientation of the robot with respect to time. On average, at the midpoint of the test, the roll angle error is 27°. While, at the end of the test, the roll angle error is 30°. Regarding the pitch angle, at the halfway point of the experiment the error is 21° and at the end of the experiment it is 19° less than the simulated state. Table 5.7 presents the results of the orientation in this experiment.

Table 5.7: Orientation results for the open-loop motion test

| | 50 % Time | | 100 % Time | |
	Roll Angle ϕ (deg)	Pitch Angle ψ (deg)	Roll Angle ϕ (deg)	Pitch Angle ψ (deg)
Simulated state IMUs	90	90	90	90
Average error back IMU	18.23	-21.33	21.98	-19.08
Average error front IMU	35.26	-21.31	37.84	-19.86
Average error IMUs	26.74	-21.32	29.91	-19.47

The robot in this experiment performed a locomotion in a straight pipeline at a constant speed. The load for this experiment is the total weight of the three modules. The error of the encoders is elevated but it is constant during the test and it can be caused by the load applied to the gearmotors. Nevertheless, the difference between the measured distance by the ruler and the encoder reading is less than 5 %. This discrepancy between readings can be caused by the lower travelled distance or by measurement errors.

Fig. 5.8 (c) shows that the pitch angle of the back propulsive module starts at 90° in some cases but it decreases rapidly after a few seconds, until the propulsive module falls on its shoulders. This is caused by the lack of a closed-loop controller with a 3D dynamic model. Fig. 5.8 (a) illustrates that the front propulsive module has a constant angle for the duration of the experiments. This is true since the front module is placed into the pipe manually at the beginning of the experiment. Thus, the bottom shoulder of this propulsive module is in contact with the pipe wall the entire time.

Fig. 5.8 (b) and (d) depict how the propulsive module twisted inside the pipe during the test. The maximum roll angle error in the robot is of almost 30° at the end of the simulation. This rotation inside the pipe is mainly triggered by the fact that the gravity centre of the robot is not centred at the beginning of the test. The initial position of the robot is with one arm at the bottom and the other three arms holding the position of the robot. To have a centred gravity centre, the orientation of the robot must be rotated by 45°. However, this position was not executed because the robot uses a dynamic controller based on a 2D model.

These results show that the robot requires a closed-loop controller that uses the three orientation angles and the angular position of the motors to adjust its position and orientation along the pipeline.

5.2.4   Closed-Loop Motion Test in a Straight Pipe

This closed-loop test evaluates the precision of the encoders and of the gearmotors when the robot uses a PD controller without the dynamic model of the robot. The PD controller uses a proportional constant (P) and a derivative constant (D) to modify the input of the gearmotors. This controller was directly coded in the on-board computer to avoid transmission delays. Fig. 5.9 depicts the PD controller for this experiment.

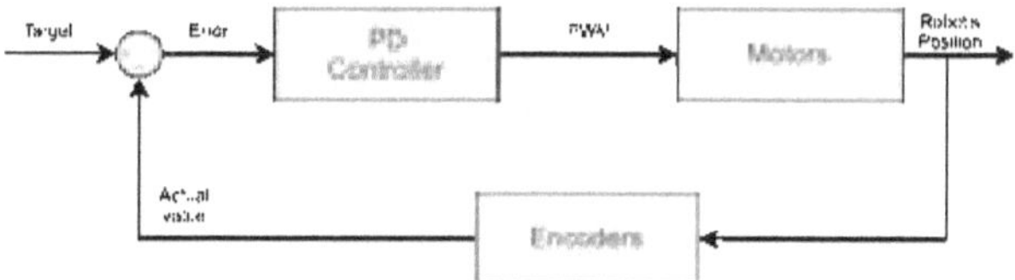

Figure 5.9: PD controller for the closed-loop motion test

The starting position of the robot, as well as the initial and final conditions, are the same as in the previous open-loop motion test. The controller target was set to reach 100 cm as fast as possible, while minimizing the error between the input and the output.

The controller estimates the position of the robot with the average reading of the eight encoders. The error between the actual value and the desired value is multiplied by a proportional and a derivative constant to calculate the required speed of the gearmotors. This velocity is converted to a single PWM signal that is sent to all the gearmotors. It is important to note that the encoders' average and the single PWM signal can only be used when the robot executes a linear motion. The test was repeated ten times and the encoder readings were plotted alongside the error. Fig. 5.10 shows the average reading of the encoders with respect to time. This figure uses markers to identify the travelled distance measured by the ruler.

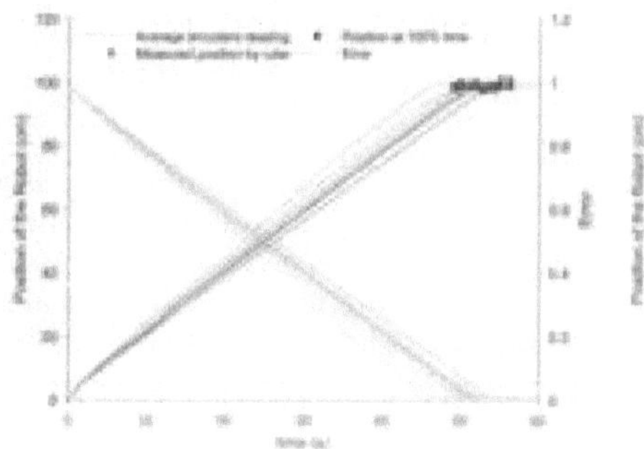

Figure 5.10: Tracked trajectories of the encoders for the closed-loop motion test

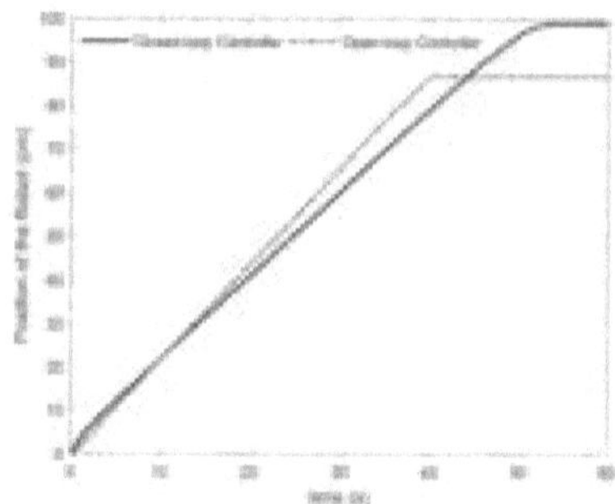

Figure 5.11: Average of open-loop trajectories and closed-loop trajectories

On average, the robot reached its final target after 52.5 seconds. The average error of the encoders at the end of the experiment was 1.2 % less than the target, while the real travelled distance measured by the ruler was 1.0 % less. The PD controller in this experiment showed promising results to reach the desired position with the encoders' average as feedback. The discrepancy between the encoders and the ruler is mainly caused by the short distance travelled and by possible errors when measuring the travelled distance with the ruler. The average speed of the robot during this experiment was 1.9 cm/s.

Fig. 5.11 depicts the average trajectory of the open-loop motion tests alongside the average trajectory of the closed-loop motion tests. This graph shows that the open-loop experiments had an average error of 13 %, while the closed-loop experiments had an average error of less than 2 %. These errors confirm that the robot needs a closed-loop controller in order to achieve the required motion inside the pipeline. It is also seen that the response of the closed-loop test is smoother than the response of the open-loop test. This outcome helps in the robot stability because it avoids violent movements on the motors.

Tuning PD Controller

A manual tuning was performed in order to select the best values of the PD controller constants. After the tuning process, the selected constants were P=0.022 and D=0.001. This approach used a single motor in a no-load environment, and it is detailed as follows:

- The proportional constant was set at P=0.001, while the derivative constant was set at D=0.000

- The proportional constant was increased by 0.001 until the motor had enough speed to reach the target in a reasonable amount of time, while still avoiding aggressive braking towards the end.

- The derivative constant was increased by 0.001 until the final response presented a minimum overshoot.

5.2.5   Endurance Test

This experiment estimates the actual autonomy of the robot when it is moving in a straight pipeline. For this test, the robot was placed inside an 8-inch pipe in the same position as in the previous experiment. The robot moved back and forth along the pipe for 100 cm. For every cycle of this test, the servomotors were driven from 0 until they reached the pipe wall. Note that the test was made with a fully charged battery of 3000 mA h at 7.4 V.

The text file created by the optimized controller for the open-loop motion test was modified to reset the position of the servomotors every time the robot returned to its initial position. The experiment was repeated ten times until the batteries could not supply enough energy to execute the routine. Table 5.8 depicts the results in this experiment.

Table 5.8: Experimental results for the endurance test

Test	Time (min)
1	29
2	38
3	31
4	37
5	30
6	37
7	28
8	36
9	30
10	37
Average	33

The robot in this experiment simulated a locomotion inside a pipeline network. It can be seen from the results that the robot lasts less time in the odd tests than in the even tests. This was caused because the robot used two different sets of batteries during the experiment. So, while one was being charged the other was being used by the robot. The capacity of the lithium-polymer battery decreases with the quantity of charge/discharge cycles. This means that the batteries used in the odd tests are older or they have been charged more times than the second set of batteries. Nevertheless, the batteries are strong enough to provide the electronics with enough power for the robot to perform at least 28 minutes of continuous motion inside the pipe.

5.3 Chapter Summary

The robot developed in this thesis was subjected to different scenarios in order to validate its components as well as its design. The encoder tests showed that the dynamic model must include the gearmotor model to reduce the error during a trajectory execution and it must use a closed-loop controller that utilizes the encoder readings as feedback. The open-loop motion test showed that the robot needs to use a 3D dynamic model and the three orientation angles of the robot as a feedback, in order to keep the equilibrium in the centreline of the pipe while following the required trajectory. The closed-loop motion test proved that the error in the encoder tests is caused by the lower distance travelled. This test also corroborated that the robot needs a closed-loop controller in order to achieve a desired position inside the pipeline.

Chapter 6

Discussion

6.1 Design/Prototype

The component dimensions caused an issue during the design process because the necessary torques can only be achieved with components that cannot fit into the robot's optimal dimensions. The selected components modified the distance between shoulder joints, which are now 15 % smaller than the optimal value. Because of this modification, the robot could not travel through 6-inch elbows with the actual controller. However, this limitation can be overcome with continuous rotation servomotors or with a more robust controller.

The robot dimensions affect not only the propulsive module but also the control module. The biggest problem in this module is that the USB and the Ethernet ports on the on-board computer are not accessible when it is mounted on the case. For this reason, a python script runs before the initialization of the on-board computer. This program searches and connects to any known Wi-Fi networks to access remotely. In case the program does not detect known networks, it creates a Hot-Spot for the user to connect any device with Wi-Fi and SSH capabilities such as computers, cellphones or tablets.

Access to the USB and to the Ethernet port can be enabled with a double-sided custom board that holds four motor drivers per side. This board can be placed in the bottom side of the control module and will allow the user to have free access to these ports.

6.2 Simulations

The actual controller is based on a 2D dynamic model that does not take into consideration the roll and pitch angle during the trajectory. For instance, the simulations were executed in a two-dimensional environment assuming that the robot does not twist inside the pipe. The simulations in elbows showed that the velocities and torques exerted by the second propulsive module are larger than the ones exerted by the first propulsive module. To reduce these differences, the pose of the controller must be modified in order to contemplate the position of the centre of mass and the orientation of both propulsive modules. This pose will ensure that the controller rectifies the position error between each centre of mass and the tracking path, and not only between the first propulsive module and the tracking path.

6.3 Experiments

The experimentation process was required to evaluate the performance of the actuators, the sensors and the on-board computer without mathematically processing the dynamic model at the same time. The main problems were the orientation of the propulsive modules and the speed of the robot.

The orientation problem was originated by the gravity centre of the propulsive modules because the gravity centre at the beginning of the tests was not centred in the pipeline. The robot has a centred gravity centre when the arms are positioned at $45°$, $135°$, $225°$ and $315°$. However, this orientation was not used because the controller uses a 2D dynamic model and it needs the arms to be at $0°$, $90°$, $180°$ and $270°$. The variation of the pitch angle is caused by the torque due to gravity and by the lack of a closed-loop controller. This, in turn, caused the propulsive module to fall off a few seconds after the test started. This event led to an increase of the friction force that the gearmotors needed to overcome, since the shoulders were scratching the inner wall of the pipe. Nevertheless, the implementation of a closed-loop controller with a 3D dynamic model that uses the three orientation angles of the robot as inputs will ensure that the robot is in equilibrium in both the horizontal and the vertical planes.

During the experimental tests inside the pipeline, the robot did not reach the required velocity given by the on-board computer because the weight of the robot reduced the speed of the gearmotors. The encoder tests also show that the gearmotors do not rotate at the same speed with the same PWM signal. In order to avoid speed differences between motors, the robot must use a closed-loop controller that independently corrects the velocity of each gearmotor with its in-built encoder.

The open-loop experiments show that the average error between the encoders and the ruler measurements was of 5%. However, the closed-loop test reduced the error to less than 2%. Although the robot did not experience significant slippage during the experiments, it can indeed slip when it executes harder maneuvers or when it moves on surfaces of different materials. If the slippage is caused by the friction between the shoulders and the wall of the pipe, it can be reduced by placing passive wheels on the spring hub of the shoulder. These wheels will allow a smooth motion in case the controller is not able to compensate the pitch angle of the propulsive modules.

The experimental tests used the internal sensors of the robot to measure the required position and orientation of the robot. However, the measurements calculated by these sensors can be biased. In order to estimate possible biases, an online algorithm such as a Kalman filter can be implemented to estimate the pose of the robot along with the measurement biases. In addition to this algorithm, the utilization of external sensors must be added to validate the encoder readings.

6.4   Closed-Loop Controller with Dynamic Model

A closed-loop controller that uses the 3D dynamic model of the robot needs to know the position of the front propulsive module with respect to the origin. Although the encoders keep track of the rotational displacement on each wheel, this sensor does not provide the real position of the propulsive module because the wheels can drift. The position of the front propulsive module tells the robot when there is an object in front or when it needs to start a steering motion. This information can be obtained from a distance sensor that uses the I2C protocol because the actual configuration of the on-board computer does not have available digital pins to read the sensor. In case the sensor does not have I2C communication, a new microprocessor must be added between the encoders and the on-board computer to free digital pins in the on-board computer. Another option to detect objects in front of the robot is to use a camera with a CSI connector that runs a background process to identify boundaries inside the pipe, allowing the robot to determine its environment. This controller must be directly coded in the on-board computer to avoid transmission delays between the robot and an external computer. This code should also contemplate the use of quaternions to detect the orientation of the propulsive modules from the IMUs.

Chapter 7

Conclusion

The objective of this work was to prototype and implement an autonomous in-pipe inspection robot for small diameter pipes. The simulations demonstrate that the robot needs active shoulder joints with independent movement in order to follow the required pathline. The experimental results prove that the robot followed the controller outputs but with some errors. However, the main errors were caused by variables that are not considered in the actual controller and by the speed variations between gearmotors. Thus, the controller must be modified to adapt the roll and pitch angle to the robot's orientation, and it must also add the encoder readings in order to individually adjust the speed of each gearmotor.

7.1   Contributions

The design of an autonomous in-pipe robot able to travel in pipelines with 6 inches to 8 inches of diameter was proposed and developed. The robot was successfully tested by simulations in straight pipes, and in elbows of 90° and 180°. The simulations proved that the robot can inspect any type of single elbow in pipes with a diameter of 7 inches to 8 inches.

The optimal radius of the wheels and the distance between shoulder joints were obtained to maximize the range of motion of the arms. They are evaluated at 1.55 cm and 5.1 cm respectively. These parameters were used to adapt the dynamic controller into the robot's design. The robot was experimentally tested in different scenarios to evaluate the precision of the sensors, the accuracy of the actuators and the autonomy of the robot. The average error between the encoders and the real distance travelled was less than 5 % for the open-loop motion and less than 2 % for the closed-loop motion. Finally, the autonomy of the robot was evaluated at between 28 and 38 minutes, when executing continuous motion.

7.2 Future Work

In continuation of the current research on developing a closed-loop trajectory tracking controller, the next step is to add more sensors such as proximity sensors and cameras to identify the exact position of the robot. The proximity sensors can be placed along the propulsive module and point to the wall of the pipe to automatically adjust the angle of the arms with respect to the diameter of the pipe. The cameras can be placed at the front and the back of the robot to know when the robot is approaching an elbow and needs to start a steering maneuver. Furthermore, the closed-loop controller with the 3D dynamic model must be directly coded into the on-board computer to avoid transmission time delays.

Cable management must also be analyzed. The actual design contemplates spaces to run cables through the propulsive and control module. However, quick connectors between modules can be implemented to avoid damages to the cables during inspections. This connector can also be useful for wire identification and for easy interchange between propulsive modules.

Lastly, the addition of passive wheels on the shoulder must be considered to decrease any friction that the shoulders can create in case the propulsive module loses control of the pitch angle of the robot during an inspection. The actual shoulder design has a spring hub in case a spring has to be added but it can also be used as a support for the passive wheel.

References

[1] Jingzhou Dai, Yishen Xu, and Wenzeng Zhang. Spc robot: A novel pipe-climbing robot with spiral extending of coupled differential. In *2017 IEEE International Conference on Robotics and Biomimetics (ROBIO)*, pages 1088–1093, December 2017.

[2] Jun Okamoto, Julio C. Adamowski, Marcos S. G. Tsuzuki, Flavio Buiochi, and Claudio S. Camerini. Autonomous system for oil pipelines inspection. *Mechatronics*, 9(7):731 – 743, 1999.

[3] Zheng Hu and Ernest Appleton. Dynamic characteristics of a novel self-drive pipeline pig. *IEEE Transactions on Robotics*, 21(5):781–789, October 2005.

[4] Zhangjun Song, Hongliang Ren, Jianwei Zhang, and Shuzhi Sam Ge. Kinematic analysis and motion control of wheeled mobile robots in cylindrical workspaces. *IEEE Transactions on Automation Science and Engineering*, 13(2):1207–1214, April 2016.

[5] Y. Kawaguchi, I. Yoshida, H. Kurumatani, T. Kikuta, and Y. Yamada. Internal pipe inspection robot. In *Proceedings of 1995 IEEE International Conference on Robotics and Automation*, volume 1, pages 857–862 vol.1, May 1995.

[6] Yoon-Gu Kim, Dong-Hwan Shin, Jeon-Il Moon, and Jinung An. Design and implementation of an optimal in-pipe navigation mechanism for a steel pipe cleaning robot. In *2011 8th International Conference on Ubiquitous Robots and Ambient Intelligence (URAI)*, pages 772–773, November 2011.

[7] Young-Sik Kwon and Byung-Ju Yi. Design and motion planning of a two-module collaborative indoor pipeline inspection robot. *IEEE Transactions on Robotics - TRob*, 28:681–696, June 2012.

[8] Se-gon Roh and Hyouk Choi. Differential-drive in-pipe robot for moving inside urban gas pipelines. *IEEE Transactions on Robotics*, 21(1):1–17, February 2005.

[9] Xiao Yu, Yonghua Chen, Michael Z. Q. Chen, and James Lam. Development of a novel in-pipe walking robot. In *2015 IEEE International Conference on Information and Automation*, pages 364–368, August 2015.

[10] Kejie Xu, Hong Xu, Qikuan Yu, Zhiqiang Wang, and Wenjiao Xu. A novel crawling in-pipe robot design. *MATEC Web of Conferences*, 61:04017, 01 2016.

[11] Yanheng Zhang, Mingwei Zhang, Hanxu Sun, and Qingxuan Jia. Motion analysis of a flexible squirm pipe robot. In *2010 International Conference on Intelligent System Design and Engineering Application*, volume 1, pages 959–963, October 2010.

[12] Atsushi Kakogawa, Taiki Nishimura, and Shugen Ma. Development of a screw drive in-pipe robot for passing through bent and branch pipes. In *International Symposium on Robotics*, pages 1–6, October 2013.

[13] P. Li, W. Yang, X. Jiang, and C. Lyu. Active screw-driven in-pipe robot for inspection. In *2017 IEEE International Conference on Unmanned Systems (ICUS)*, pages 608–613, Oct 2017.

[14] B. Klaassen and K. L. Paap. Gmd-snake2: a snake-like robot driven by wheels and a method for motion control. In *Proceedings 1999 IEEE International Conference on Robotics and Automation (Cat. No.99CH36288C)*, volume 4, pages 3014–3019 vol.4, May 1999.

[15] Hagen Schempf and George Vradis. Explorer: Untethered real-time gas main assessment robot system. In Ger Maas and Frans Van Gassel, editors, *Proceedings of the 20th International Symposium on Automation and Robotics in Construction ISARC 2003 – The Future Site*, pages 471–476, September 2003.

[16] Christopher J. Fisher. Using an accelerometer fon inclination sensing. *Convergence Promotions*, May 2011. [Online]. Available: https://www.digikey.com/en/articles/techzone/2011/may/using-an-accelerometer-for-inclination-sensing. [Accessed: 28-Dec-2019].

[17] Lounis Douadi, Davide Spinello, and Wail Gueaieb. Dynamics and control of a planar multibody mobile robot for confined environment inspection. *Journal of Computational and Nonlinear Dynamics*, 10:011005, September 2014.

[18] Extending the life of aging pipeline infrastructure. [Online]. Available: https://www.gptindustries.com/en/downloads/extending-life-aging-pipeline-infrastructure. [Accessed: 11-Jan-2020].

[19] Andrew Reeves and J.t Ryan. Pipelines in canada. *The Canadian Encyclopedia*, June 2019.

[20] Hyouk Choi and Se-gon Roh. *In-pipe Robot with Active Steering Capability for Moving Inside of Pipelines*, chapter 23. IntechOpen, September 2007.

[21] Muhammad Azri Abdul Wahed and Mohd Rizal Arshad. Wall-press type pipe inspection robot. In *2017 IEEE 2nd International Conference on Automatic Control and Intelligent Systems (I2CACIS)*, pages 185–190, October 2017.

[22] H.R. Choi and S.M. Ryew. Robotic system with active steering capability for internal inspection of urban gas pipelines. *Mechatronics*, 12(5):713 – 736, 2002.

[23] Glen Bright and Devon Ferreira. Automated pipe inspection robot. *Industrial Robot*, 24(4):285–289, 1997.

[24] Tokuji Okada and Takeo Kanade. A three-wheeled self-adjusting vehicle in a pipe, ferret-1. *The International Journal of Robotics Research*, 6(4):60–75, 1987.

[25] K. Suzumori, T. Miyagawa, M. Kimura, and Y. Hasegawa. Micro inspection robot for 1-in pipes. *IEEE/ASME Transactions on Mechatronics*, 4(3):286–292, Sep. 1999.

[26] M. Muramatsu, N. Namiki, R. Koyama, and Y. Suga. Autonomous mobile robot in pipe for piping operations. In *Proceedings. 2000 IEEE/RSJ International Conference on Intelligent Robots and Systems (IROS 2000) (Cat. No.00CH37113)*, volume 3, pages 2166–2171 vol.3, Oct 2000.

[27] Iszmir Nazmi Ismail, Adzly Anuar, Khairul Salleh Sahari, Mohd Zafri Baharuddin, Muhammad Fairuz, Abd Jalal, and Juniza Md Saad. Development of in-pipe inspection robot: A review. In *2012 IEEE Conference on Sustainable Utilization and Development in Engineering and Technology (STUDENT)*, pages 310–315, October 2012.

[28] A. Kakogawa and S. Ma. Mobility of an in-pipe robot with screw drive mechanism inside curved pipes. In *2010 IEEE International Conference on Robotics and Biomimetics*, pages 1530–1535, Dec 2010.

[29] T. Li, S. Ma, B. Li, M. Wang, and Y. Wang. Design of spring parameters for a screw drive in-pipe robot based on energy consumption model. In *Proceeding of the 11th World Congress on Intelligent Control and Automation*, pages 3292–3297, June 2014.

[30] P. Li, S. Ma, B. Li, Y. Wang, and Y. Liu. Self-rescue mechanism for screw drive in-pipe robots. In *2010 IEEE/RSJ International Conference on Intelligent Robots and Systems*, pages 2843–2849, Oct 2010.

[31] Ren Tao, Liu Qingyou, Li Yujia, and Chen Yonghua. Basic characteristics of a novel in-pipe helical drive robot. *Int. J. of Mechatronics and Automation*, 4:127 – 136, January 2014.

[32] Kundong Wang, Wencan Gao, and Shugen Ma. Snake-like robot with fusion gait for high environmental adaptability: Design, modeling, and experiment. *Applied Sciences*, 7:1133, November 2017.

[33] G. Robinson and J. B C Davies. Continuum robots - a state of the art. In *Proceedings of the 1999 IEEE International Conference on Robotics and Automation, ICRA99*, volume 4, pages 2849–2854, 1999.

[34] Lounis Douadi, Davide Spinello, Wail Gueaieb, and Hassan Sarfraz. Planar kinematics analysis of a snake-like robot. *Robotica*, 32:659–675, August 2013.

[35] Jeff Lamonde. Development of a multi-body autonomous inspection robot for small diameter pipes. Master's thesis, University of Ottawa, 2017.

[36] E. Oberg, F.D. Jones, H.L. Horton, H.H. Ryffel, C.J. McCauley, R.M. Heald, and M.I. Hussain. *Machinery's Handbook: A Reference Book for the Mechanical Engineer, Designer, Manufacturing Engineer, Draftsman, Toolmaker, and Machinist.* Machinery's Handbook: A Reference Book for the Mechanical Engineer, Designer, Manufacturing Engineer, Draftsman, Toolmaker, and Machinist. Industrial Press, 2004.

[37] Danielle Collins. Worm gears: What are they and where are they used, Oct 2017. [Online]. Available: https://www.motioncontroltips.com/worm-gears-what-are-they-and-where-are-they-used/ [Accessed: 02-Feb-2020].

[38] Tianyuan Yang. Structural design and optimization of multi-body autonomous inspection robot for small diameter pipelines. resreport, University of Ottawa, 2018.

[39] IBT Inc. All electric motor are not created equal, Aug 2012. [Online]. Available: https://www.ibtinc.com/electric-motors-are-not-created-equal/ [Accessed: 5-Jul-2020].

[40] Rolled Alloys. Pipe size chart, Feb 2016. [Online]. Available: https://www.rolledalloys.ca/tools/pipe-chart/ [Accessed: 10-Apr-2020].

Appendix A

Standard Pipe Dimensions

The Nominal Pipe Size (NPS) in Canada is standardized by the norm ASME B36.10M and B36.19M. Every pipe is identified by the outside diameter, the nominal pipe size, and the schedule, which is a non-dimensional number that represents the nominal wall thickness of the pipe. Table A.1 shows the dimensions in inches for 6-inch and 8-inch pipes for 40S & standard schedule.

Table A.1: Pipe dimensions for standard schedule

Nominal Pipe Size	Outside Diameter	Wall Thickness	Inside Diameter
6	6.625	0.280	6.065
7*	7.625	0.301	7.023
8	8.625	0.322	7.981

The pipe dimensions of the fittings are standardized by the norm ASME B16.9. The elbows and the straight pipes use the same centreline radius. There are two types of radius included in this norm, the long radius and the short radius. The long radius is 1.5 times the pipe diameter while the short radius is 1.0 times the pipe diameter. Table A.2 shows the dimensions in inches for 6-inch and 8-inch pipes for 40S & standard schedule.

Table A.2: Elbows dimensions for standard schedule

NPS	Outside Diameter	Wall Thickness	Inside Diameter	Short Radius	Long Radius
6	6.625	0.280	6.065	6	8
7*	7.625	0.301	7.023	7	10.5
8	8.625	0.322	7.981	8	12

*The pipe dimensions for 7-inch are not included in the norm ASME. However, some companies such as Rolled Alloys [10] in Canada provide this type of pipe.

Appendix B

Normal Forces and Wheel Torques

The minimum normal forces described in Section 3.2.1 were obtained with the following
Matlab code. This code uses the actual values of the propulsive module to estimate the
most accurate forces.

```
clc
clear all
close all
%Initial parameters
z        = 0.03;
armL     = 0.116;
h        = (0.150)+(z*2);
A        = 0.65 ;
linkW    = 0.051;
H        = h*(A-0.5) ;
Lb       = A*h ;
Lf       = (1-A)*h ;
rhoW     = 0.0155;
inch2m   = 0.0254;
pipeW    = 6.065*inch2m;
m1       = 0.78*2;
m2       = 1.12*1.25;
m3       = m2;
u        = 0.4;
g        = 9.81;
%Complementary distances
army = (pipeW/2)-(linkW/2)-(rhoW);
armx = sqrt((armL.^2)-(army.^2));
%Effective distances
lfg2 = armx-H;
lfg1 = armx-H+Lf;
lff1 = pipeW;
```

```matlab
%Alpha angle
alphaarm = asin(army/armL);
%Gravitational forces
Fg2 = m2*g;
Fg1 = m1*g/2;
Fg3 = m3*g/2;
%Maximum normal forces in the horizontal pipe scenario
Fn1 = ((Fg2*lfg2)+(Fg1*lfg1))/(u*lff1)
Fn2 = Fn1-Fg1-Fg2
%Maximum normal forces in the vertical pipe scenario
Fnv= (Fg2+Fg1+Fg3)/(4*u)
%Wheel torque
Tw = (rhoW/2)*(Fg1+Fg2+Fg3)
```

Appendix C

Range of Motion in the Arms

In this section, the range of motion of the arms for straight pipelines with 6 inches to 8 inches diameter is obtained. This range represents the robot adaptability to varying pipe diameters. Fig. C.1 shows the kinematic model of the robot when there is no rotation in a straight pipeline.

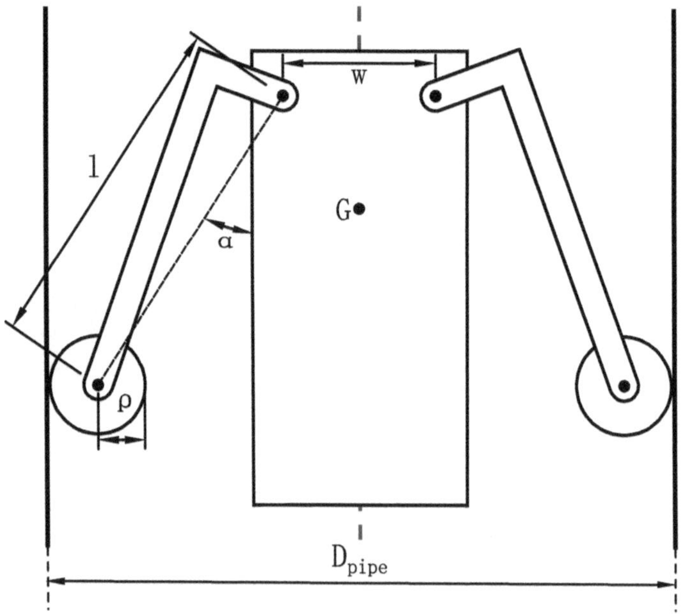

Figure C.1: Kinematic model of the robot to obtain the range of motion of the arms

Considering that the gravity centre of the robot is in the centreline of the pipe and knowing the length of the arm l, the wheel radius ρ, the distance between arm joints w, and the inner pipe diameter D_{pipe}, the following equation can be obtained:

$$D_{pipe} = 2\rho + w + 2l \sin \alpha \tag{C.1}$$

Thus, the alpha angle can be estimated by:

$$\alpha = \arcsin \left(\frac{D_{pipe} - 2\rho - w}{2l} \right)$$

The angle for a 6-inch pipe is 18.1° and the angle for an 8-inch pipe is 31.35°. Therefore, the necessary range of motion of the arms for straight pipeline ascends to 13.25°.

Appendix D

Mechanical Range of the Arms

The minimum and maximum arm angle that the robot can reach is mechanically limited by the worm screws plates. Fig. D.1 illustrates the robot in Solidworks and it shows one arm in its minimal arm position which is 10.74°. Thus, the minimal angle of the arm is set at 11.00°.

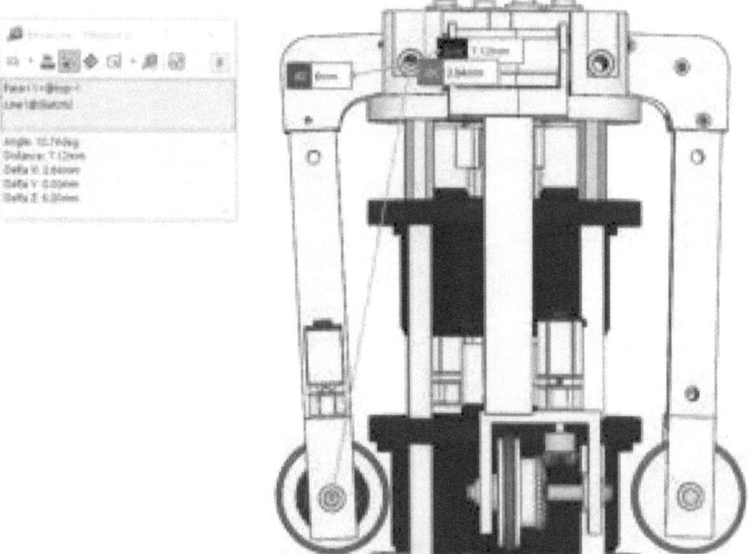

Figure D.1: Minimum mechanical angle of the arm

Fig. D.2 shows the robot with one lifted arm at its maximum arm angle which is 43.20°. Then, the maximum angle of the arm is set at 43.00°.

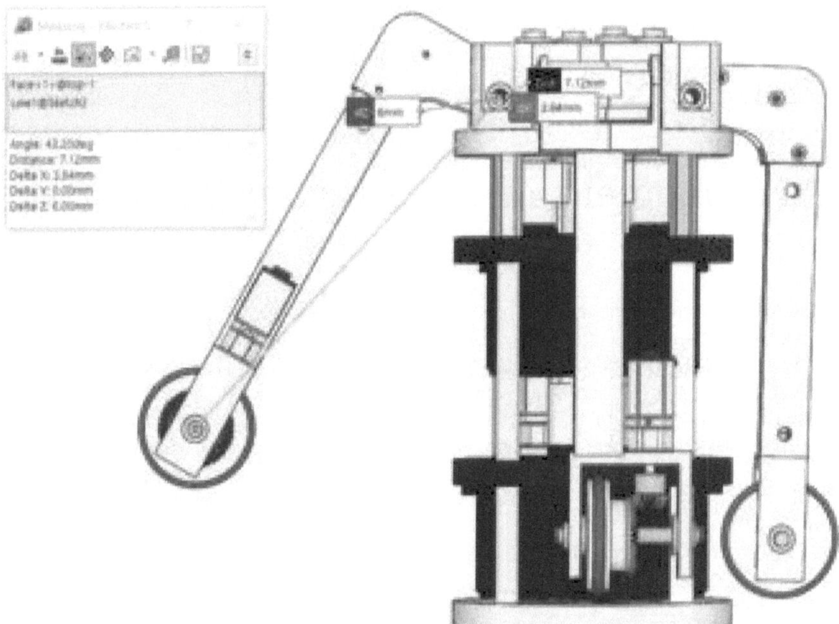

Figure D.2: Maximum mechanical angle of the arm

Appendix E

Arrangement of the Servomotors

The internal components of the propulsive module were designed according to the selected servomotor dimensions. As the propulsive module needs four arms, there is a need of four servomotors, one per worm drive mechanism. Fig. E.1 depicts a diagram that shows four servomotors inside a circle with a diameter of 73.70 mm because it is the maximum body diameter minus a wall of 1.25 mm. The distance between servomotor must be higher than 6 mm, which is the worm shaft diameter because two servomotors must be placed at the bottom and the worm shafts go through the body.

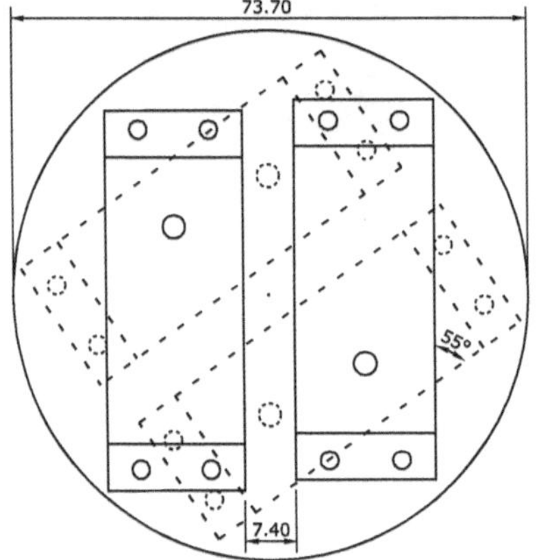

Figure E.1: Servos plate dimensions

Figure E.2: Worm screw upper plate

Knowing the distances between servomotors, a plate for the worm screws can be designed. The most important design parameter for the propulsive module is the distance between active arm joints, which its optimal value is 60 mm. Fig. E.2 illustrates the final plate design, which is symmetric about an axis for easy machining and assembling. This plate has nine holes at the top, four for bushings where the worm screws rotate, two for any cable running through the body, and one for a bolt to fix the universal joint.

Appendix F

Stresses in the Arm

In this section, the maximum exerted stresses on the arm are estimated to find if the arm beam could be deflected. Fig. F.1 illustrates the forces on the aluminum profile between the wheel holder and the arm shoulder. In this configuration, the arm beam will be directly affected by the forces generated in the bevel gears and the normal force calculated in Section 3.2.1.

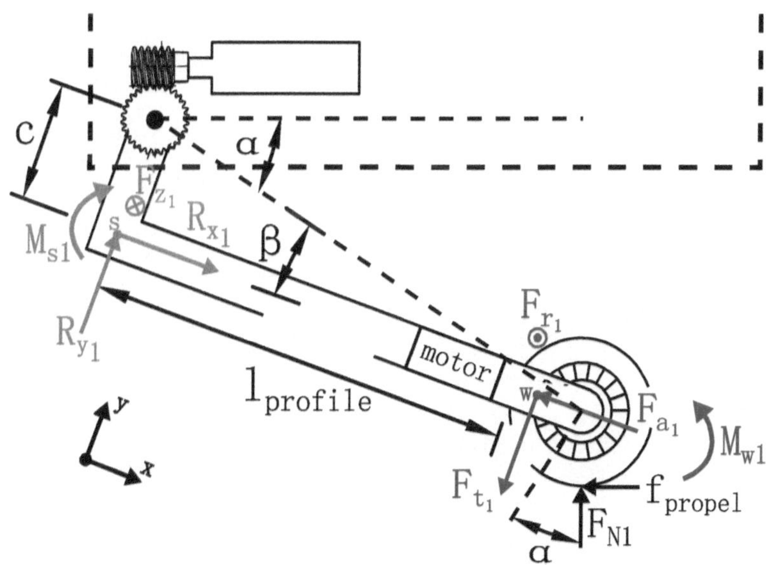

Figure F.1: Forces on the aluminum profile of the arms

The following equations represent the forces in the bevel gear mechanism:

$$F_{t1} = \frac{T_{motor}}{5\, d_{pinion}} \tag{F.1}$$

$$F_{r1} = F_{t1} \tan\gamma \cos\delta \tag{F.2}$$

$$F_{a1} = F_{t1} \tan\gamma \sin\delta \tag{F.3}$$

where $\gamma = 20°$ because the bevel gears are straight and $\delta = \tan^{-1}\frac{z_1}{z_2} = \tan^{-1}\frac{20}{40} = 26.56°$

Since all the parameters are known, we can calculate the forces in the bevel gear mechanism as follows:

$$F_{t1} = \frac{0.8}{5 * 0.01} = 16.00\,\text{N}$$

$$F_{r1} = 16 * \tan 20 * \cos 26.56 = 5.20\,\text{N}$$

$$F_{a1} = 16 * \tan 20 * \sin 26.56 = 2.60\,\text{N}$$

The propel force generated by the gearmotor in the wheel can be calculated as:

$$f_{propel} = \frac{n_{bevel}\, \eta_{bevel}\, T_{motor}\, \eta_{motor}}{r} \tag{F.4}$$

$$f_{propel} = \frac{2 * 0.95 * 0.8 * 0.9}{0.016} = 85.50\,\text{N}$$

The reaction forces in the arm shoulder are not collinear to the normal and the propel forces. Thus, we need to calculate the angle to rotate these forces. This angle can be estimated with the angles alpha and beta. Alpha is a variable that depends on the robot's motion and beta can be obtained with the measurement of c and $l_{profile}$ with the following equation:

$$\beta = \text{atan}\frac{c}{l_{profile}} = \text{atan}\frac{27.43}{112.75} = 13.67°$$

The reaction forces and the moments are calculated in two planes, xy and xz plane. The principal components are F_{t1}, F_{a1}, f_{propel}, and F_{N1} . The equations for the reactions can be obtained with the sum of the forces in each axis or plane.

For x-axis:

$$\sum F_x = 0 \tag{F.5}$$

$$R_{x1} = -F_{a1} + f_{propel}\cos(\alpha - \beta) + F_{N1}\sin(\alpha - \beta) \tag{F.6}$$

$$R_{x1} = -2.6 + 85.5 * \cos(\alpha - 13.67°) + 36.1 * \sin(\alpha - 13.67°)$$

For y-axis:

$$\sum F_y = 0 \tag{F.7}$$

$$R_{y1} = F_{t1} - f_{propel}\sin(\alpha - \beta) + F_{N1}\cos(\alpha - \beta) \tag{F.8}$$

$$R_{y1} = 16 - 85.5 * \sin(\alpha - 13.67°) + 36.1 * \cos(\alpha - 13.67°)$$

For z-axis:

$$\sum F_z = 0 \tag{F.9}$$

$$R_{z1} = -F_{r1} \tag{F.10}$$

$$R_{z1} = -5.2\,\text{N}$$

For xy plane:

$$\sum M_{xy} = 0 \tag{F.11}$$

$$M_{w1} = F_{N1}\, r_{driven} - f_{propel}\, r_{wheel} \tag{F.12}$$

$$M_{w1} = 36.1 * 0.01 - 85.5 * 0.016 = -1.03\,\text{N m} \tag{F.13}$$

$$M_{s1} = -(M_{w1} + R_{y1}\, l_{profile}) \tag{F.14}$$

$$M_{s1} = 1.03 - 0.112 * (16 - 85.5 * \sin(\alpha - 13.67°) + 36.1 * \cos(\alpha - 13.67°))$$

For xz plane:

$$\sum M_{xz} = 0 \tag{F.15}$$

$$M_{s2} = -F_{r1}\, l_{profile} \tag{F.16}$$

$$M_{s2} = -5.2 * 0.112 = -0.58\,\text{N m}$$

These forces and moments can be applied together to obtain the shear and bending stresses. With the intention of get these stresses in a hollow beam, the inertia of the cross section has to be obtained. Fig. F.2 illustrates the sectional view of the aluminum beam of the arms.

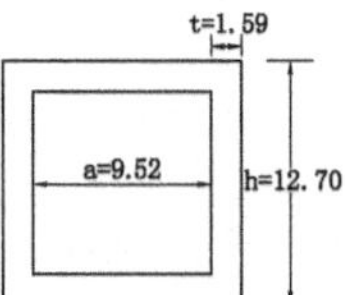

Figure F.2: Hollow aluminum beam cross section

The inertia of a the hollow beam, can be found from:

$$I_{profile} = \frac{h^4}{12} - \frac{a^4}{12} \tag{F.17}$$

$$I_{profile} = \frac{0.0127^4}{12} - \frac{0.00952^4}{12} = 1.4819 \times 10^{-9} m^4$$

Von Misses stress is important to determine if a specific material will yield due to forces or moments. This equation is mostly used in ductile material, such as aluminum. The bending stresses are defined with the following equations:

$$\sigma_{11} = \frac{M_{s1} \times y_{max}}{I_{profile}} \tag{F.18}$$

$$\sigma_{11} = \frac{(1.03 - 0.112 * (16 - 85.5 * \sin(\alpha - 13.67°) + 33 * \cos(\alpha - 13.67°))) \times 0.00635}{1.4819 \times 10^{-9}}$$

$$\sigma_{22} = \frac{M_{s2} \times y_{max}}{I_{profile}} \tag{F.19}$$

$$\sigma_{22} = \frac{-0.58 \times 0.00635}{1.4819 \times 10^{-9}} = 25.17 \, \text{MPa}$$

The shear stresses are describes as:

$$\sigma_{12} = \frac{R_{y1} \times S_n}{b \, I_{profile}} \tag{F.20}$$

where $S_n = (2\,t\,h\,\Delta_x) + \left(4 \times \left(\frac{a}{2} \times t \times \Delta_y\right)\right) = 26.4 \times 10^{-6} m^4$

$$\sigma_{12} = \frac{16 - 85.5 * \sin(\alpha - 13.67°) + 33 * \cos(\alpha - 13.67°) * 26.4 \times 10^{-6}}{0.0112 * 1.4819 \times 10^{-9}}$$

$$\sigma_{23} = \frac{R_{z1} \times S_n}{b\,I_{profile}} \tag{F.21}$$

$$\sigma_{23} = \frac{-5.2 * 26.4 \times 10^{-6}}{0.0112 * 1.4819 \times 10^{-9}} = -730.92\,\text{kPa}$$

Therefore, the equivalent Von Misses stress can be found from:

$$\sigma_s = \sqrt{\frac{1}{2}\left[(\sigma_{11} - \sigma_{22})^2 + (\sigma_{22} - 0)^2 + (0 - \sigma_{11})^2 + 6 \times (\sigma_{12}^2 + \sigma_{23}^2 + 0^2)\right]} \tag{F.22}$$

The previous results are dependent of α which is the arm's angle. This value varies from 22° to 43° for a motion in an 8-inch pipeline. The maximum equivalent stress is 120 MPa and the yield strength for aluminum is 241 MPa. Fig. F.3 illustrates the results for the range of theta.

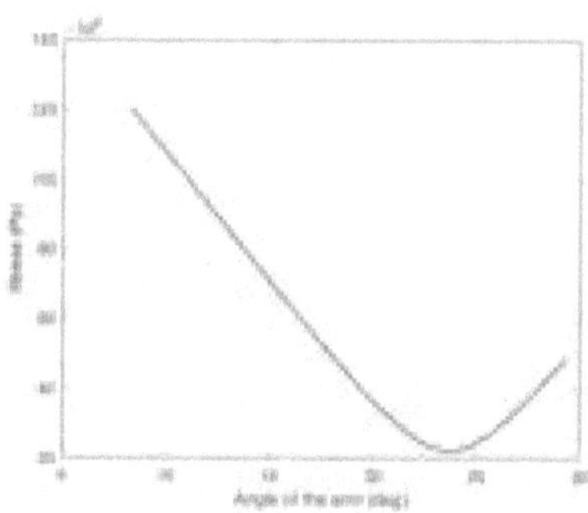

Figure F.3: Equivalent stress in the hollow arm beam

Appendix G

Motor Drivers and Encoders

In this section, the custom boards of the control module are analyzed. These boards gather the motor drivers and the encoders. Fig. G.1 shows the Pinout of a motor driver DRV8838 and Fig. G.2 shows the pinout of the encoder TLE4946-2K.

Figure G.1: Pinout motor driver

Figure G.2: Pinout encoder

As can be seen in Table G.1 and G.2 , the power voltage varies between 0 V and 18 V depending of the device and the purpose. However, the voltage for the actuators and the encoders was standardized to 6 V. This modification reduced the quantity of wires and the connection time of the components. Each encoder has six pins and the pins were divided in two cables of three wires each. The first cable (Vcc, M2, M1) goes directly to the motor drive output (Vcc, Out2, Out1). The second cable goes to another section that does not have directly contact with the motor driver because those wires are the sensor readings and they go directly to the on-board computer.

Table G.1: Pins description for encoders

Component	Pin	Description	Value	Type
	GND	Ground	0 V	Power
	Out B	Channel B	N/A	Output
Encoder	Out A	Channel A	N/A	Output
	Vcc	Logic Voltage	2.7-18 V	Power
	M2	Terminal 2 Motor	N/A	Input
	M1	Terminal 1 Motor	N/A	Input

Table G.2: Pins description for motor driver

Component	Pin	Description	Value	Type
	GND	Ground	0 V	Power
	Vcc	Logic Voltage	1.8-7 V	Power
	Vin	Motor Voltage	0-11 V	Power
	Enable	Motor Speed	PWM	Input
Motor driver	Phase	Motor Direction	Logic 1/0	Input
	Out2	Terminal 2 Motor	N/A	Output
	Out1	Terminal 1 Motor	N/A	Output
	Sleep	Sleep Mode (Boolean)	Logic 1/0	Input
	Vm	Motor Voltage	0-11 V	Power

Appendix H

Actuators Tests

In this section, an estimation of the current consumption of the actuators is developed. This problem was sundered in various tests that the robot executed. For these tests, the robot was controlled through the GUI and it was placed outside the pipe in an orientation where the wheels do no generate any propulsion. There are two types of tests, the first one varies the speed of all the gearmotors from 10% to 100%, while the second one varies the speed of the gearmotors and also move the servomotors from 0° to 180°.

The actuator signals were sent directly from the external computer to the on-board computer. A multimeter was added between the batteries and the voltage regulators to measure the continuous current of the robot during the test. Each test was repeated ten times at different speeds. Fig. H.1 depicts the continuous current that the robot demanded depending on the gearmotors speed. On average the robot demands 1.9 A when it uses the eight gearmotors, while it demands 3.5 A when it uses all the gearmotor and servomotors at the same time. Considering that the batteries supply 7.4 V at the beginning of each set of test, the power consumption of the robot ascends to 14.06 W when the robot uses the eight gearmotors, while it consumes 25.90 W when it controls the gearmotors and the servomotors at the same time.

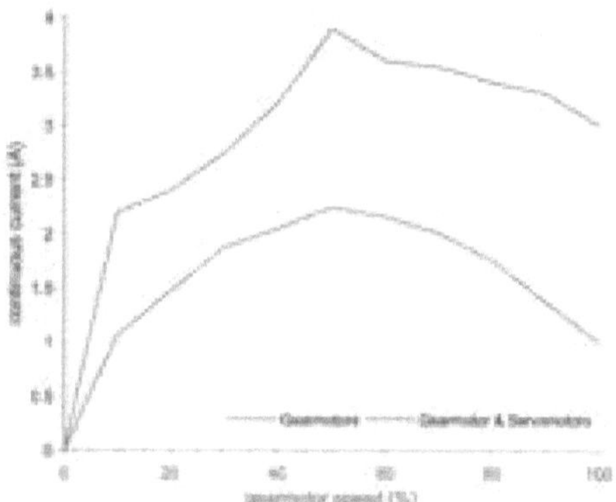

Figure H.1: Current consumption of the robot at different speeds of the gearmotors

The results from this problem suggest that the robot must avoid the use of medium speeds on the gearmotors when it is controlling the position of the servomotors. Fig. H.1 shows that the current peak is presented at 50 % of the speed of the gearmotors.

Surprisingly, the current did not increase proportionally with the speed of the gearmotor. This effect can be caused by the other electronics that are been used to activate the gearmotors, such as motor drivers and Servo HAT. Nevertheless, it can also be caused by the absence of load on the gearmotors.

Appendix I

User Manual

In this section, an user manual is presented. It starts with the architecture description of the robot and its components description. Later, it describes how to establish wireless communication to the robot with the two connection modes. The next section explains how to adjust the angle of the arms and servomotors. Lastly, the graphical user interface (GUI) is detailed as well as the use of it.

I.1   Robot Architecture

The presented robot shown in Fig. I.1 is comprised of three modules: two propulsive modules and one control module. The propulsive modules have four active joints for the rotational displacement of each arm, and every arm has a wheel at the end to make contact on the inner wall of the pipe. The control module has an on-board computer, a set of batteries, and electronics.

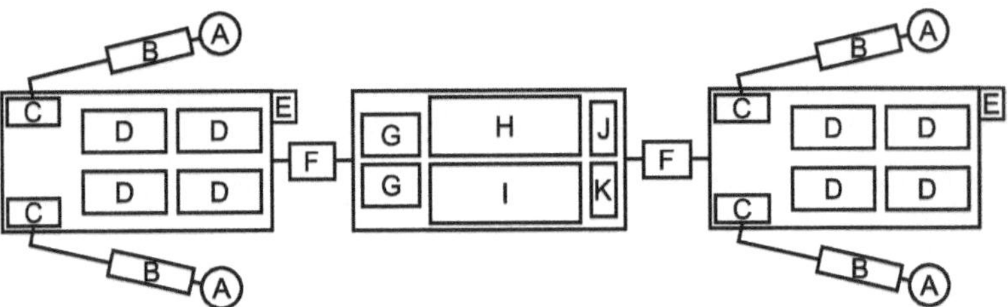

Figure I.1: Architecture of the robot

Labels from Fig.I.1 are detailed in Table I.1

Table I.1: Components description of the robot

Label	Component	Description	Product Code Brand
A1	Wheel	$\varnothing$=31mm	N/A
A2	Bevel gear mechanism	$\varnothing_{shaft}$ =4 mm Reduction ratio, 1:2	A 1B 3MYK05020 A 1B 3MYK05040
B	Gearmotor	Stall Torque, 0.8 Nm Reduction ratio, 1000:1	Pololu
C	Worm gear mechanism	$\varnothing_{shaft}$ =6 mm Reduction ratio, 1:10	AG0.5-20R2/KWG0.5-R2
D	Digital Servomotor	Stall Torque, 1.5 Nm Voltage, 6V	EzRobot Heavy-Duty
E	IMU	MPU6050, 6 DOF	DFRobot
F	Universal Joint	$\varnothing_{shaft}$ =3/8" Max. angle, 45°	McMaster, 60625K95
G	Motor Driver	DRV8838 Max. Current, 1.7A	Pololu
H1	Raspberry pi	Model, 3 b+ Raspbian Buster	Raspberry Pi
H2	Servo HAT	PCA9685, 16 Channels	Adafruit Industries
I	Batteries	1500 mA 2S 25-50C lithium-polymer	Turnigy nano-tech
J	Voltage converter	D24V150F6, 6V / 20A	Pololu
K	Voltage converter	Lm22676, 5V / 3A	Phi Robotics

I.2 Wireless Connection to the Robot

The robot needs two programs in the external computer to work: an X11 display server, and terminal emulator that supports SSH and X11 protocols. This manual was developed using Xming and PuTTY in the external computer.

The robot has two wireless connection modes: Hotspot mode and router mode, both of them use the Wi-Fi of the on-board computer. The hotspot mode is a direct connection from the external computer to the on-board computer, while the router mode is an indirect connection, since it uses a router as a bridge between the on-board computer and the external computer.

The Raspberry Pi has a script that runs during the initialization for wireless connection purposes. This program searches and connects to any known Wi-Fi networks to access remotely. In case the program does not detect known networks, it creates a Hot-Spot for the user to connect any device with Wi-Fi and SSH capabilities such as computers, cellphones or tablets.

Hotspot Mode

This connection gives the user a complete control over the robot through the GUI. If the external computer is connected using this mode, it will not have internet connection unless it is connected to a LAN network.

The Wi-Fi Hotspot has the following settings:

$$SSID = PipeRobot$$
$$Password = uOttawa123$$

Follow these steps to connect the external computer to the robot:

1. Search for networks

2. Connect to the robot's SSID

3. Enter the SSID's password

Once the connection has been established, the robot and the external computer must be in the same network segment. The default IP address of the robot in Hotspot mode is "192.168.50.5". To verify that the connection is successful, open the windows command processor (CMD) and write, 'ping 192.168.50.5', it will show the time response between the external computer and the robot like in Fig. I.2.

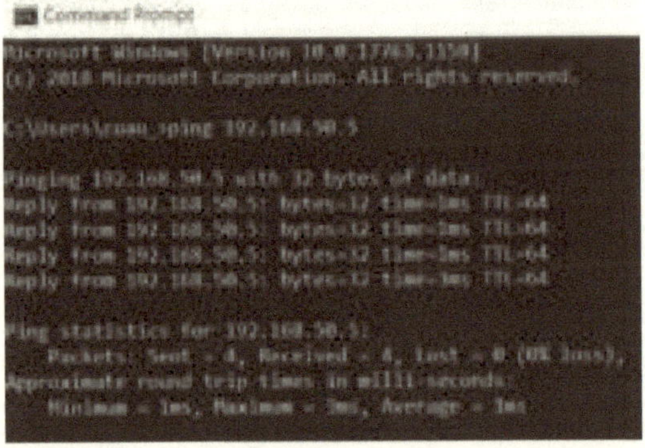

Figure I.2: Verification of the robot connection

Router Mode

This mode provides the same Hotspot benefits and it also gives access to the desktop view of the on-board computer. This type of connection allows the external computer to have access to internet through the Wi-Fi.

The on-board Wi-Fi settings are:

$$MAC\ Address = B8{:}27{:}EB{:}2A{:}7C{:}E9$$
$$Device\ Name = PipeRobot$$

The Raspberry Pi configures the Wi-Fi networks by DHCP, which means that the router grants a specific IP address to the robot. In order to establish a connection to the robot in this mode, the IP address must be known. The easiest way to find this address is accessing to the router and following the next steps:

1. Access to the router settings

2. Go to 'Association List' or 'Devices List'

3. Search the device with the same MAC address and device name

4. Save the IP address that is displayed on the list

Fig. I.3 shows a sample of the devices list in the router settings, where the IP address in this case is "192.168.0.135".

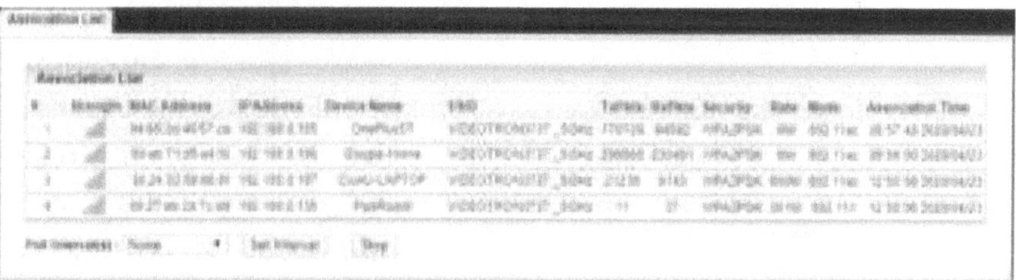

Figure I.3: Devices list sample

To verify that the IP address is correct, a ping command with the on-board IP address can be sent. This procedure is shown in the previous section 'Hotspot Mode'.

Command Line of the Robot

The robot needs a terminal emulator to enter in the robot's command line and execute the GUI program or any other program. In order to have a good connection with the robot, follow the next steps to setup PuTTY:

1. Install PuTTY in the external computer

2. Open PuTTY

3. In the category three on the left side, click in the "+" next to SSH

4. Select X11

5. Enable X11 forwarding like in Fig. I.4

6. Click on Session in the category tree

7. Change the port and the connection type like in Fig. I.5

8. In Host name enter the IP Address of the robot that depends on the connection mode

9. Click on Open

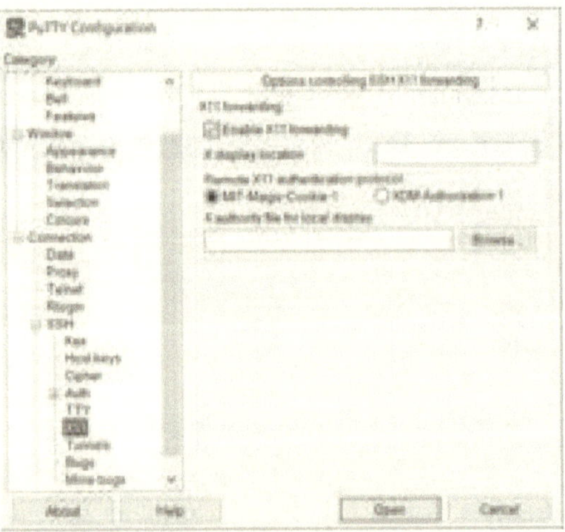

Figure I.4: X11 configuration

The emulator will prompt if the configuration was correct. This window is the robot's command line and it will ask for the robot credentials which are:

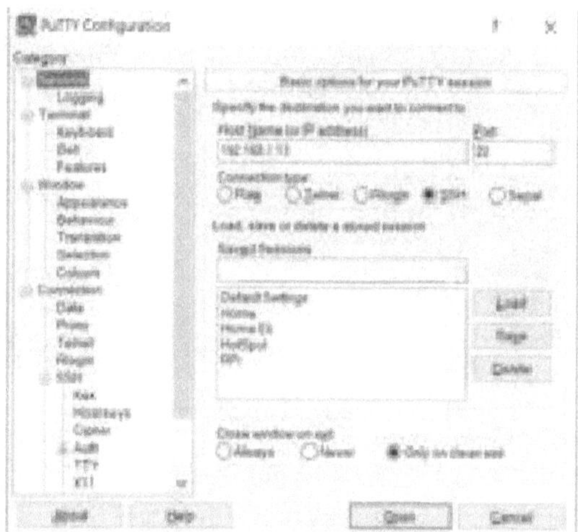

Figure I.5: SSH configuration

user : pi
password : raspberry

Fig. I.6 shows the command line when the connection is successful.

Add a New Wi-Fi Network

The addition of a new Wi-Fi network in both connection modes is made with the robot's command line. Before adding a new network, write down the credentials and follow the next steps:

1. Open the robot's command line (See connection modes)

2. Open the network configuration file, with the following command:

```
sudo nano /etc/wpa_supplicant.conf
```

3. Go to the end of the file and add the credentials of the new Wi-Fi network with the following format:

```
network= {
    ssid="Network_Name"
    psk= "Password"
}
```

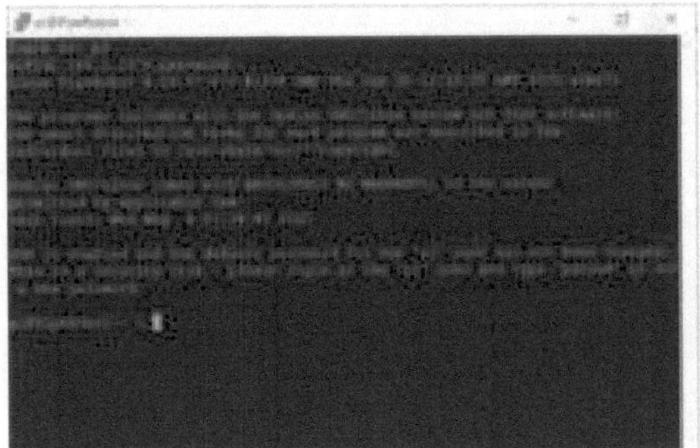

Figure I.6: Robot's command line

4. Using the keyboard, press 'Ctrl+x', then 'y', and 'enter'

5. Restart the Raspberry Pi, with the next command:

```
sudo reboot
```

6. Wait 5 seconds

7. Turn on and turn off the robot with the general switch

I.3   Adjusting the Angle of the Arms

The angle of the arms must be adjusted, if the initial angle of the arms is different than 11°. If an adjustment is necessary, the first step is to loose the aluminum couples (using a 7/64" allen key) that are between the worm drive mechanisms and the servomotors. Fig. I.7 depicts four red circles that shows the location of the four couples that each propulsive module has.

Each arm is moved by a worm drive mechanism and the angle can be manually adjusted when moving the worm screws with a screwdriver. These worm screws have a slot that is facing the upper plate of the propulsive module as can be seen in Fig. I.8. The angle of the arm decreases when moving the worm screw in a clockwise direction and it increases when moving the worm screw in an anti-clockwise direction.

When the angle of the arm is adjusted in the correct position. The last step is to adjust the position of the servomotors to the required servomotor angle (normally is 0°) with the user interface. Finally, using the allen key thigh the servomotor couples.

Figure I.7: Side of a propulsive module showing the servomotor couples

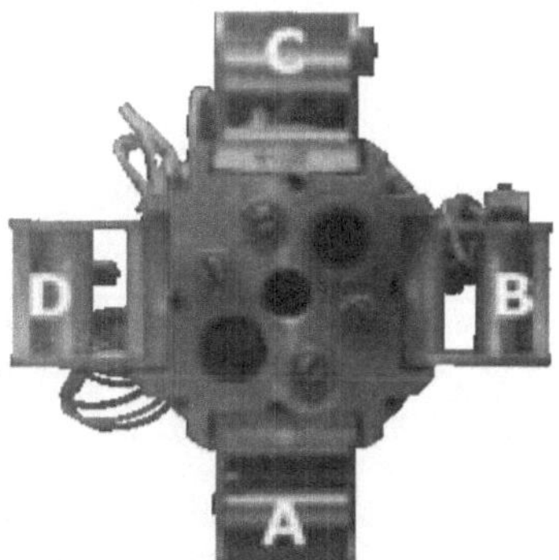

Figure I.8: Propulsive module with labelling

I.4 User Interface

All the propulsive modules have a specific label in each arm. Fig. I.8 shows a propulsive module seen from the top view, where the arm at the bottom is called arm "A". This representation is used to know what is the initial orientation of the robot inside the pipeline and it is also used for reference on the user interface.

A GUI was designed to send and receive data between the Raspberry Pi 3 B+ and the external computer. Fig. I.9 illustrates the main view of the interface. The GUI script is constantly reading the sensor values to calculate the inclination angles of the propulsive modules and the angular position of the wheels. It also sends the necessary data to the drivers when a slider bar, button or check button changes state.

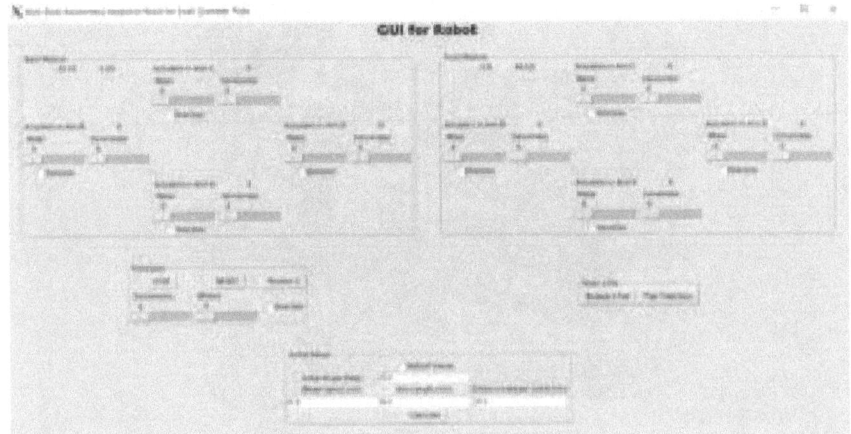

Figure I.9: GUI to control the robot

The interface is divided in five sections and each section is represented by a square frame with a title in the top left corner. The sections at the top all have the same functionalities, expect that the left section controls the propulsive module at the back and the right section controls the propulsive module at the front. These two sections have eight slide bars, four to change the speed of the gearmotors from 0 to 32 rpm and the others to change the position of the servomotors from 0° to 180°. They also contain check buttons to individually change the rotation on the gearmotors; the robot moves forward by default.

The section in the middle-left is called "Functions" and it contains pre-programmed routines that the robot can follow, as well as general functions to control all of the servomotors and gearmotors. These functions stop the rotation of the gearmotors, reset the position of the servomotors, move the servomotors to a specific position or start the gearmotors at a specific speed.

The section in the middle-right is called "Open file" and it has three buttons. The "Browse a file" button allows the user to open the file manager and search for a text file that contains the robot's routine. The "Play Trajectory" button reads the selected text file and sends the speed or position of the actuators. Lastly, the "Stop Trajectory" button stops the execution of the selected text file.

The last section at the bottom is called "Initial Setup" and it helps the user initialize the angle of the arms for the inspection. The check button at the top is to choose the actual dimension of the robot such as wheel radius, arm length and distance between shoulder joints. In case the user is using a robot with different dimensions, these values can be inputted manually. The initial angle field is given in the text file generated by the dynamic controller. The "Calculate" button displays a pop-up message with the distance between wheels. This distance must be adjusted after the servomotors were set at 0°.